Mechanics of Machines

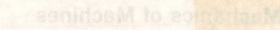

Mechanics of Machines

G. H. Ryder
and
M. D. Bennett

Second Edition

Industrial Press Inc.
200 Madison Ave., New York, NY 10016

Library of Congress Cataloging-in-Publication Data
Ryder, G. H. (Geoffrey Harwood)
Mechanics of machines/G.H. Ryder and M.D. Bennett.–2nd ed.
p.   cm.
ISBN 0–8311–3030–X
1. Mechanical engineering.   I. Bennett, M. D. (Michael David)
II. Title.
TJ146.R9   1990
621.8′11–dc20
                                                           90-34523
                                                           CIP

Published in Great Britain 1990 by
MACMILLAN EDUCATION LTD
London and Basingstoke

Printed in Hong Kong

Industrial Press
200 Madison Avenue
New York, New York 10016

Mechanics of Machines

Contents

Preface to the First Edition

The present text has evolved and grown from an earlier book by one of the authors entitled *Theory of Machines through Worked Examples*. When it was decided that the time had come for conversion to SI units, the opportunity was taken to change the format from a collection of problems to a conventional student-textbook. In the event, this has enabled a more comprehensive and self-contained work to be produced and the title has been modified to reflect this change. The book has been completely rewritten under co-authorship and this has provided an opportunity to incorporate fresh ideas in teaching and presentation of the subject.

The book begins with a chapter on the fundamental principles of mechanics of machines and, although this is necessarily extensive, a conscious effort has been made to deal only with those aspects which are essential to the understanding of later work. Throughout the book the aim has been to illustrate the theory by means of engineering applications over a wide field. Numerical examples have been carefully selected to give the reader a realistic appreciation of actual system parameters and performance.

The chapters then follow a traditional sequence, dealing first with the kinematics of machines, and with force and velocity relations in systems where inertia effects may be neglected. Subsequently, attention is turned to the analysis of specific machine elements such as clutches, belt drives and gears. This is followed by consideration of inertia effects in machines, leading naturally to the problems of balancing rotating and reciprocating masses. Three chapters are devoted to vibrations, both free and forced, of single- and multi-degree-of-freedom systems. The book ends with a chapter on the principles and applications of control in mechanical engineering.

Preface to the Second Edition

Recent changes in the structure of mathematics and physics courses in universities, and the depth of coverage given, have indicated that some complementary changes to the *Mechanics of Machines* text were appropriate. The main differences between the first and second editions occur in the sections dealing with the basic principles of mechanics, which now appear as Part I and are arranged in six chapters instead of the original single chapter. This allows the individual principles to be more readily identified and systematically studied. At the same time, more worked examples have been included, and some topics have been developed a little further, notably in the new chapter on impulse and momentum.

Part II contains the original chapters that described some of the applications of the principles discussed in Part I. Again, the opportunity has been taken to add to some of these chapters. A section on strength and wear has been included with toothed gears, cams have been included as a further example of machine elements, and engine balance has been extended to Vee-engine layouts. The final chapter on control has been enlarged to include the application of control systems to robotics.

Each chapter in the second edition concludes with a summary of the important results for easy reference.

While making these changes, care has been taken not to change the general approach, and the expected readership is the same.

PART I: PRINCIPLES

1 Introduction

1.1 SCOPE OF THE SUBJECT

Mechanics is that science which deals with the action of forces on bodies and of the effects they produce. It is logically divided into two parts—*statics* which concerns the equilibrium of bodies at rest, and *dynamics* which concerns the motion of bodies. Dynamics, in turn, can be divided into *kinematics*— which is the study of the motion of bodies without reference to the forces which cause the motion—and *kinetics*, which relates the forces and the resulting motion.

The definition of a *machine* will be considered in greater detail in chapter 7, but in general terms it can be said to consist of a combination of bodies connected in such a manner as to enable the transmission of forces and motion. Thus the behaviour and performance of machines can be analysed by application of the principles of mechanics.

The history of statics goes back over 2000 years, the earliest recorded writings being those of Archimedes who formulated the principles of the lever and of buoyancy. Compared with statics, dynamics is a relatively new subject. The first significant contribution was made about 500 years ago by Galileo who studied the fall of bodies under gravity. About 100 years later, Newton, guided by the work of Galileo and others, put dynamics on a sound basis by the accurate formulation of the *laws of motion*.

Chapters 1 to 6 cover the basic principles of statics, kinematics and kinetics, necessary for an understanding of the applications to be studied in subsequent chapters.

1.2 MATHEMATICAL MODELS

The solution of engineering problems depends on certain physical laws and uses mathematics to establish the relationships between the various quantities involved. The ability to make the transition between the *physical system* and

3

the *mathematical model* describing it is vital to the solution of a problem. However, it is important to remember that the model gives only an ideal description, that is, one that approximates to, but never completely represents, the real system.

To construct the mathematical model for any problem, certain approximations and simplifications are necessary, depending on the nature of the problem, the accuracy and extent of the data, the analytical tools at one's disposal, and the accuracy required in the answer. For example, consider the earth. For problems in astronomy, the earth may be treated as a particle, while for predicting eclipses it may be considered as a rigid sphere. For work on earthquakes the elastic properties of the earth must be taken into account. In all the problems and examples in following chapters, the earth has been assumed to be infinitely large and flat.

In practice, only the major or the less obvious assumptions are actually stated but the reader is strongly advised to consider very carefully the implied assumptions in every worked example and problem. Setting up a good mathematical model is the most important single step and many difficulties stem from a failure to do this.

1.3 SCALARS AND VECTORS

Mechanics deals with two kinds of quantities. *Scalar quantities* such as time, mass and energy have magnitude but no directional properties, and can be handled by the ordinary laws of algebra. *Vector quantities*, however, are associated with direction as well as with magnitude, and include force, displacement and acceleration.

Vector quantities can be represented graphically by straight lines with arrowheads (sometimes called directed line-segments), such as V_1 and V_2 in figure 1.1a, and are printed in bold type to distinguish them from scalars. The length of each line represents the magnitude of the respective quantity, while the angles θ_1 and θ_2 define their directions relative to some convenient datum.

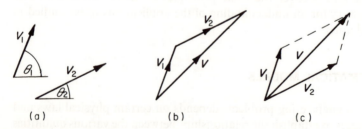

(a)   (b)   (c)

Figure 1.1   Addition of two vectors

Vectors are of three kinds, *fixed*, *sliding* and *free*, and an example of each is now given.

(a) If a force is applied to a deformable body, the effect will generally be dependent on the point of application, as well as on the magnitude and direction. Such a force can only be represented by a fixed vector, that is, one fixed in space.
(b) On the other hand, a force applied to a rigid body can be considered to act at any point along its line of action without altering its overall effect, and can thus be represented by a sliding vector.
(c) If, however, a rigid body moves without rotation, then all points have the same motion, and so the motion of the body as a whole can be completely described by the motion of any point. We could thus choose any one of a set of parallel vectors to represent this motion. Such a vector is described as free.

1.3.1 Vector Addition and Subtraction

Sliding and free vectors can be added by the triangle or parallelogram laws as shown in figure 1.1. The combined vector, V, called the *resultant*, of the two vectors V_1 and V_2 is written as

$$V = V_1 + V_2$$

By drawing the vectors V_1 and V_2 nose to tail as shown in figure 1.1b, the sum V can be obtained by joining the free ends and adding the arrowhead such that the path followed by V leads to the same point as the path followed by taking V_1 and V_2 consecutively. Figure 1.1c shows that the sum is independent of the order in which the vectors are added.

If the vector V_2 is drawn in the opposite direction, we have the vector $-V_2$. The vector difference

$$V' = V_1 - V_2$$

is then easily obtained as shown in figure 1.2.

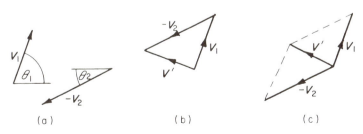

Figure 1.2   Subtraction of vectors

Where there are more than two vectors, the sum may be obtained by successive application of the triangle law, figure 1.3.

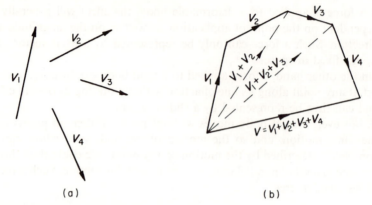

(a)   (b)

Figure 1.3   Addition of a system of vectors

In any particular case, the result of a vector addition or subtraction operation may be determined by scale drawing and measurement, or by solving the vector diagram geometrically. Geometric solutions have the potential for greater accuracy, but it is strongly recommended that a sketch, approximately to scale, always be drawn as a check against gross errors.

Example 1.1

Vectors P and Q have magnitudes of 7 and 13 units, and are at angles of 30° and 140° respectively relative to some datum. Find the magnitude and direction of each of the vectors $R = P + Q$ and $R' = P - Q$, first by scale drawing and measurement, and then by using geometry.

By measurement from figure 1.4a

$$R = P + Q = 12 \text{ units}$$

and

$$\theta = 108°$$

By geometry, the internal angle between P and Q is

$$30 + (180 - 140) = 70°$$

Using the cosine rule

$$R^2 = P^2 + Q^2 - 2PQ \cos 70°$$

hence

$$R = \sqrt{(7^2 + 13^2 - 2 \times 7 \times 13 \cos 70°)} = 12.48 \text{ units}$$

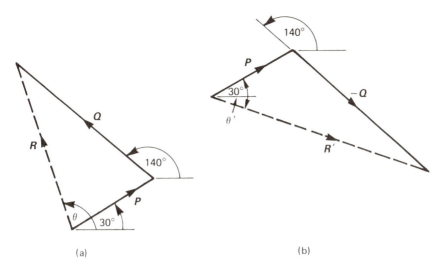

Figure 1.4

Using the sine rule

$$Q/\sin(\theta - 30°) = R/\sin 70°$$

hence

$$\theta = 30° + \sin^{-1}(13 \sin 70°/12.48) = 108.2°$$

In a similar manner, by measurement from figure 1.4b

$$R' = P - Q = 17 \text{ units}$$

and

$$\theta' = 343° \text{ or } -17°$$

while by geometry

$$R' = \sqrt{(7^2 + 13^2 - 2 \times 7 \times 13 \cos 110°)} = 16.74 \text{ units}$$

and

$$Q/\sin(\theta' + 30°) = R'/\sin 110°$$

from which

$$\theta' = 16.86°$$

Note that the calculation gives a solution for θ' corresponding to the *principal* value. It is only from examination of figure 1.4b that the correct solution, namely $-16.86°$ or $343.14°$, can be deduced. The importance of sketching a scale diagram should now be clear.

1.3.2 Resolution of Vectors

It is frequently useful to be able to split up a vector, or *resolve* it into two parts. The directions of the two parts are usually, though not necessarily, perpendicular to each other. In figure 1.5 the vector V is shown resolved into components along the x- and y-axes to give V_x and V_y respectively, where

$$V = V_x + V_y$$

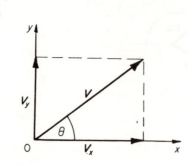

Figure 1.5   Resolution of vectors

Alternatively, the magnitudes of V_x and V_y can be expressed as

$$V_x = V \cos \theta$$
$$V_y = V \sin \theta$$

so that

$$V = \sqrt{(V_x^2 + V_y^2)} \tag{1.1}$$

and

$$\theta = \tan^{-1}(V_y/V_x) \tag{1.2}$$

Vector addition and subtraction can often be simplified by first resolving the individual vectors into scalar components, adding or subtracting as appropriate, and recombining to find the resultant.

Example 1.2

Rework example 1.1 (figure 1.4) by first resolving P and Q into x and y components.

$$P_x = 7 \cos 30° = 6.06, \qquad Q_x = 13 \cos 140° = -9.96$$

hence

$$R_x = 6.06 + (-9.96) = -3.90 \text{ units}$$

and

$$P_y = 7 \sin 30° = 3.50, \qquad Q_y = 13 \sin 140° = 8.36$$

hence

$$R_y = 3.50 + 8.36 = 11.86 \text{ units}$$

Combining

$$R = P + Q = \sqrt{(R_x^2 + R_y^2)} = \sqrt{[(-3.90)^2 + 11.86^2]} = 12.48 \text{ units}$$

and

$$\theta = \tan^{-1}(R_y/R_x) = \tan^{-1}(-11.86/3.90) = -71.8°$$

Once again, the correct direction must be deduced from the principal value of $-71.8°$ and a sketch of the solution, figure 1.6a.

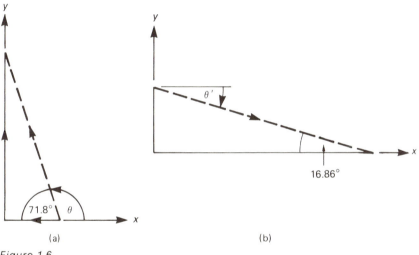

(a)   (b)

Figure 1.6

In a similar manner

$$R_x' = P_x - Q_x = 6.06 - (-9.96) = 16.02$$

and

$$R_y' = P_y - Q_y = 3.50 - 8.36 = -4.86$$

so that

$$R' = P - Q = \sqrt{(R_x'^2 + R_y'^2)} = 16.74 \text{ units}$$

and

$$\theta' = \tan^{-1}(-4.86/16.02) = -16.86°$$

Figure 1.6b shows that the principal value gives the required answer directly.

1.4 FUNDAMENTAL QUANTITIES

The whole of mechanics requires just three fundamental quantities, namely mass, length and time, commonly referred to as dimensions. All other quantities in mechanics such as acceleration, force, energy and momentum, can be derived from these three fundamental quantities.

It is important to note that the description of physical quantities requires both the magnitude and the units to be specified, and, in the case of vectors, also the direction. Units will be discussed in more detail in chapter 2.

It is good practice to check equations for dimensions, and also for units if possible; errors can often be trapped with this technique.

1.5 PROBLEMS

1. Three vectors have magnitudes of 7, 12 and 4 units, and make angles of 30°, 152° and −72° respectively relative to some reference direction. Find the sum of the three vectors.
   [6.27 units at 121.7°]

2. A vector of magnitude 41 units at 64° is added to a second vector of magnitude of 29 units at 312°. What third vector must be added so that the sum is zero?
   [40.4 units at 202.3°]

3. Vectors *A*, *B*, and *C* are as follows and sum to zero:

	Magnitude	Direction
A		0°
B	12	150°
C	8	

   Find the missing values in the table. Explain graphically why there are two answers.
   [*A*: 5.10 and 15.68 units, *C*: 311.4° and 228.6°]

4. A ship travels 1200 km north followed by 850 km east. Treating each of the two parts of the voyage as a vector, determine the final position of the ship from its starting point.
   What important assumption has been made about the model of the earth? How could the model be improved, and what effect would it have on the answer?
   [1471 km, N35.3°E]

2 Forces and Equilibrium

2.1 THE NATURE OF FORCES

A *force* can be considered as an interaction between two material objects. Physics recognises three such interactions, namely, gravity forces, electric forces and nuclear forces.

Gravity forces are generally only significant when at least one of the objects is large; in this text we shall only consider the particular gravity force called *weight*, which is discussed in more detail in section 2.1.3. Gravity forces act throughout the entire object or body and are described as *body forces*.

Electric forces comprise those due to magnetic effects which, in some respects, resemble gravitational effects and are also body forces, and those due to electrostatic effects, that is, interactions between objects carrying electric charges. Electrostatic effects can be demonstrated by rubbing a glass rod with animal fur and using the rod to pick up small pieces of paper. At a sub-atomic level, electrostatic forces exist between electrons and protons. These forces can attract or repel the particles according to the nature of the charges, but the magnitudes are generally only significant when the spacing between them is small. When two objects are brought into close proximity, the electrostatic force is normally repulsive and may be quite large. In engineering terms such a condition is described as *contact*, and the interaction is described as a *surface force*. We shall make extensive use of surface forces throughout the text.

Nuclear forces exist between all pairs of sub-atomic particles, and are rather complex in nature. However, the subject matter of the text does not require them to be taken into account.

All forces have magnitude and direction, and must therefore be manipulated according to the rules for vectors described in chapter 1.

2.1.1 Newton's Laws of Motion

Suitably reworded to meet the needs of the following chapters, Newton's laws are

First Law   A particle remains at rest, or continues to move in a straight line with constant velocity, if there is no resultant force acting on it.

Second Law   If a resultant force acts on a particle, then the particle will accelerate in the direction of the force at a rate proportional to the magnitude of the force.

Third Law   The forces of action and reaction between contacting bodies are equal, opposite and collinear.

The first law defines the condition known as *equilibrium*, and is discussed further in section 2.3.

The second law forms the basis of most of the analysis in kinetics, and is used extensively in chapter 4 and subsequent chapters. As applied to a particle of constant mass m, the second law may be stated as

$$R = kma \tag{2.1}$$

where R is the resultant force acting on the particle, a is the acceleration produced and k is a constant that will depend on the units chosen for m, R and a (see section 2.1.2). The second law can also be considered to contain the first since, if there is no acceleration, by equation 2.1 there can be no resultant force.

The third law is fundamental to an understanding of the concept of force, and is important in the construction of *free-body diagrams* (section 2.2.1).

2.1.2   Units

In equation 2.1 it is convenient to choose a coherent or consistent set of units, that is, one which makes $k = 1$. Equation 2.1 then reduces to

$$R = ma \tag{2.2}$$

Of the three quantities, force, mass and acceleration, the units of any two can be chosen arbitrarily and used to determine the units of the third. A number of such systems have evolved, as listed in table 2.1, and in each case, *one unit of force equals one unit of mass multiplied by one unit of acceleration.*

Table 2.1

| | | *Units* | |
System	*Force*	*Mass*	*Acceleration*
1	dyne	gram	centimetre/second^2
2	poundal	pound mass	foot/second^2
3	pound force	slug	foot/second^2
4	newton	kilogram	metre/second^2

In system 1 the units are small and thus give rise to numbers that are too large for convenient use in engineering. Systems 2 and 3 are widely used in the US and are still occasionally used in the UK. System 4 has been agreed internationally; it is used widely in the UK and almost exclusively in the rest of Europe. A major advantage of this system is the ease with which mechanical and electrical quantities can be related. For further details, see BS 3763:1964 The International System (SI) Units. SI units are used exclusively in this text.

2.1.3 Weight and Centre of Gravity

An example of a body force is obtained by the application of equation 2.2 to a freely falling body, which gives

$$W = m \, g$$

where W is the *weight* or gravitational attraction which the earth has for the body, and g is the resulting acceleration, about 9.81 m/s^2 at the earth's surface. Although this force is distributed over all the particles in the body, it may be assumed to be concentrated at one point as far as its overall effect is concerned. This point is known as the *centre of gravity* and may be defined as the point about which the first moment of weight is zero. In two dimensions, the co-ordinates of the centre of gravity, G, are

$$x_G = \frac{\Sigma m_i g_i x_i}{m \, g} \qquad y_G = \frac{\Sigma m_i g_i y_i}{m \, g}$$

as in figure 2.1, where the subscript i refers to a typical particle, and the summation covers all such particles.

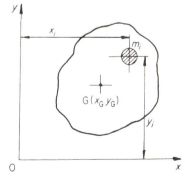

Figure 2.1 Centre of gravity

For bodies that are small compared to the size of the earth, the variation of the gravitational acceleration is negligible so that the co-ordinates of the centre of gravity become

$$x_G = \frac{\Sigma m_i x_i}{m} \qquad y_G = \frac{\Sigma m_i y_i}{m}$$

These co-ordinates actually define the *centre of mass*, that is, the point about which the first moment of mass is zero. Thus for all practical purposes, the difference between the centre of gravity and the centre of mass is rather academic.

Example 2.1

A flywheel can be considered as a solid circular disc 280 mm in diameter and with a mean thickness of 40 mm. Because of small variations in thickness, the mass centre of the flywheel is 0.5 mm from the geometric centre. To remove this eccentricity a 20 mm diameter hole is to be drilled radially into the edge of the flywheel. What depth of hole *h* should be drilled, and where should it be in relation to the mass centre?

The hole can be treated as a negative mass, and it is intuitively obvious that it should be drilled on a diameter through the original mass centre G and the geometric centre O, as shown in figure 2.2. Assuming the flywheel to be homogeneous, its mass is proportional to volume.

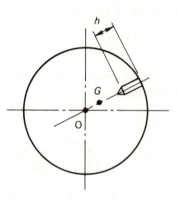

Figure 2.2

Considering first moments of mass about the final mass centre O, and ignoring the conical end of the hole

$$280^2 \times 40 \times 0.5 \times \pi/4 + (-20^2 \times h \times \pi/4)(140 - h/2) = 0$$

which gives the quadratic in *h*

$$h^2 - 280h + 7840h = 0$$

with the two real solutions 32 mm and 248 mm.

Some thought will show that both solutions would be effective, but the deeper hole is not a practical solution, and thus a 32 mm deep hole should be drilled.

2.2   SYSTEMS OF FORCES

In most problems there will be a number of forces which must be taken into account. Determining and describing such a *system of forces* is most conveniently done by means of *free-body diagrams*, which are discussed in section 2.2.1. The importance of the technique as a first step in the solution of kinetics problems cannot be overstressed. Once the free-body diagram has been drawn, the system of forces thus described can be reduced to a more simple system, as discussed in sections 2.2.2 and 2.2.3, for later use in this and subsequent chapters.

2.2.1   Free-body Diagrams

The term *free-body diagram* is used to describe the system of forces acting on a body when considered in isolation. These forces will consist of surface forces acting at points where previously there was contact with another body, together with the body forces (generally gravity). Examples for the piston and connecting rod of an IC engine are shown in figures 2.3a and b.

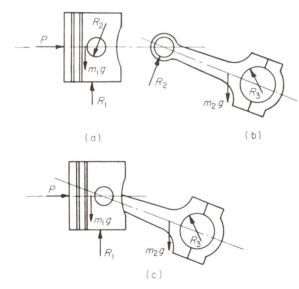

(a)   (b)

(c)

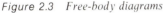

Figure 2.3   *Free-body diagrams*

The system is assumed to be in a vertical plane, so that the gravitational forces or weights mg are downwards and act at the respective centres of gravity. The pressure on the face of the piston has been assumed uniform and normal so that it can be represented by a single force P acting along the centre-line. The effect of the contact pressure between the side of the piston and the cylinder wall has also been assumed to be normal and is shown as a single force R_1 and assumed to be upwards. The direction assumed is not important provided it is taken into account when the forces (which are vectors) are added. In a similar manner the resultants of the bearing pressures R_2 and R_3, at the little and big ends respectively, are shown in assumed arbitrary directions. In any particular case, the free-body diagram, with the various assumptions outlined above, provide the mathematical model for analysis of the problem.

The technique of free-body diagrams may often conveniently be extended to deal with a group of connected bodies that are to be treated as a system in isolation. Figure 2.3c shows such a diagram for a piston and connecting rod taken together, and the forces R_2 originally shown acting at the little end no longer appear since they are now *internal* to the system.

It is important to note that, for each body or group of connected bodies, the forces shown are those that are exerted *on* the body *by* other bodies. Note, for example, the direction of the force R_2 shown on the piston and on the little end.

2.2.2   Moment of a Force

An important idea that arises in many problems is the moment of a force about some specified point or axis. Referring to figure 2.4, the moment, M, of the force F about a point O in the same plane as F, is defined as

$$M = F d$$

where d is the perpendicular distance from O onto F. Like force, moment is a vector, but in two-dimensional problems, the direction of the moment vector can be described completely by its sense, that is, clockwise or anticlockwise (the latter in this case). Summation of moments in a single plane then reduces to the algebraic sum, where a + is assigned arbitrarily to one sense or the other.

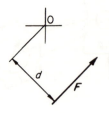

Figure 2.4   Moment of a force

If it is necessary to retain the vector concept, the *right-hand-screw rule* is normally used. When turned clockwise, a right-handed screw will advance when viewed from the head end, so that a clockwise rotational quantity is represented by a vector normal to and into the plane of rotation. In a similar manner, an anticlockwise quantity is represented by a vector projecting normally out of the plane of rotation.

2.2.3   Resultant of a System of Forces

Figure 2.5a shows an arbitrary body subjected to a number of forces F_1, F_2, F_3. The combined effect, or *resultant*, R, of these three forces is determined by the vector sum, so that

$$R = F_1 + F_2 + F_3$$

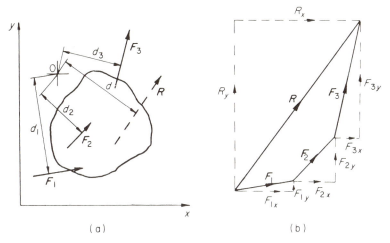

Figure 2.5   *Resultant of a system of forces*

This can be evaluated graphically by scale drawing, or analytically by resolving each force into components, say x and y, summing the components in each direction to give

$$R_x = F_{1x} + F_{2x} + F_{3x}$$
$$R_y = F_{1y} + F_{2y} + F_{3y}$$

and then recombining by the method of equations 2.1 and 2.2. Figure 2.5b shows clearly that the two methods are equivalent but, whichever is used, only the magnitude and direction of the resultant are given.

The line of action can be obtained by equating the sum of the moments of F_1, F_2 and F_3 about some arbitrarily chosen point O, to the moment of the resultant, R, about O (Varignon's principle). Taking anticlockwise positive, this gives

$$M_O = F_1 d_1 + F_2 d_2 + F_3 d_3 = Rd$$

where d is the perpendicular distance of the line of action of R from O. Hence the position of R is determined (figure 2.5a).

In some cases the vector diagram, figure 2.5b, may close by itself, so that $R = 0$; however, this does not necessarily mean that the sum of the moments is also zero. Consider the special case of two equal and opposite forces, F, distance d apart, as shown in figure 2.6. Clearly there is no resultant force, but taking moments about O, anticlockwise positive, gives

$$M_O = F(d + l) - Fl = Fd$$

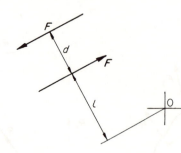

Figure 2.6   Couples

A system such as this is known as a *couple*. Note that the moment of a couple is the same about every point in its plane.

Example 2.2

Figure 2.7a shows a wireless mast to which are attached two wires exerting forces on the top, B, as shown. What should be the tension in the guy AB if the resultant force at B is to be vertical? What is the value of this vertical force?

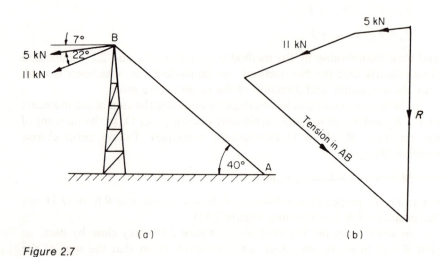

Figure 2.7

Figure 2.7b shows the vector sum of the three forces at point B. The two wire tensions are known completely but only the line of action of the guy tension is known. However, since the resultant of the three forces is to be vertical, the length of the guy tension vector can be fixed. Measuring directly from figure 2.7b

guy tension = 23.6 kN

resultant R = 22.8 kN

Example 2.3

In figure 2.8a, if the angle α is 30°, and the force and couple are to be replaced by a single force applied at a point located either (a) on line AB, or (b) on line CD, find, in each case, the distance from the centre O to the point of application of the force.

What should be the value of α if the single force is to pass through B?

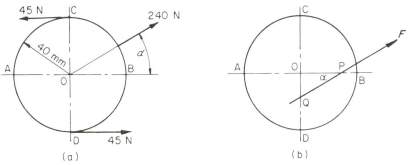

(a)   (b)

Figure 2.8

Figure 2.8b shows the general position of a single force F to satisfy parts (a) and (b). By summing forces it is clear that F equals 240 N and is at the same angle α to the horizontal.

Equating moments about O, anticlockwise positive

$$45 \times 2 \times 40 = F\, OP \sin 30° = F\, OQ \cos 30°$$

hence for (a)

OP = 30 mm

and for (b)

OQ = 17.3 mm

For F to pass through B, a moment sum about O gives

$$45 \times 2 \times 40 = F\, OB \sin \alpha$$

therefore

$\alpha = 22°$

2.3   FORCES IN EQUILIBRIUM

When the resultant of a system of forces acting on a body is zero, and when the sum of the moments of the forces about some point is also zero, then the body is said to be in *equilibrium*, that is, it will remain at rest or continue to move in a straight line with a constant speed. This is a direct consequence of Newton's first law.

In reality, all problems are three-dimensional, but in a great many cases, a two-dimensional mathematical model is justified. Equilibrium in two dimensions will therefore be considered first in some detail, followed by a brief look at equilibrium in three dimensions.

2.3.1   Equilibrium in Two Dimensions

In figure 2.5, the system of forces F_1, F_2, F_3 would be in equilibrium if

$$F_1 + F_2 + F_3 = R = 0$$

and

$$F_1 d_1 + F_2 d_2 + F_3 d_3 = M_O = 0$$

The first equation may often be more conveniently expressed in scalar components

$$F_{1x} + F_{2x} + F_{3x} = R_x = 0$$

$$F_{1y} + F_{2y} + F_{3y} = R_y = 0$$

In general, for any number of co-planar forces, equilibrium exists if either

$$\Sigma F = 0 \tag{2.3}$$

or

$$\Sigma F_x = 0 \quad \text{and} \quad \Sigma F_y = 0 \tag{2.4}$$

together with

$$\Sigma M_O = 0 \tag{2.5}$$

where O is some arbitrary point.

Application of equation 2.3 or 2.4, together with equation 2.5, entails a summation of forces, either vectorially or in mutually perpendicular directions, and a summation of moments. In some problems the solution is more easily obtained by using one of the following two alternative sets of equations which also satisfy the conditions for equilibrium.

(a) *One Force and Two Moment Equations*

If OX and OY are two reference axes in the plane, and the force summation in one direction (say OY) is zero, then if a finite resultant exists it must be parallel to OX. If, secondly, a moment sum about an arbitrary point A is also zero, then this resultant must in addition pass through A. Finally, if a further moment sum about another point B (where AB is not parallel to OX) is zero, then the resultant must be zero and the conditions for equilibrium are satisfied, that is

$$\Sigma F_y = \Sigma M_A = \Sigma M_B = 0$$

(b) *Three Moment Equations*

For the second alternative consider a moment sum about any three points A, B and C, *that are not collinear*. It is clear that all three summations can only be zero if the resultant is zero, that is

$$\Sigma M_A = \Sigma M_B = \Sigma M_C = 0$$

It is important to note that the equations contained in the alternative conditions for equilibrium are not in addition to equations 2.3, 2.4 and 2.5. Only three independent equations are necessary to define equilibrium in two dimensions, and hence only problems containing up to three unknowns are soluble, or *statically determinate*. However, sensible choice of method, and selection of the point or points about which to take moments, will often give a simpler and more elegant solution.

Special Cases of Equilibrium

(a) Two forces alone in equilibrium must be equal, opposite and collinear, that is, act in the same straight line.

(b) Three forces alone in equilibrium are concurrent, that is they must pass through a common point. This is easily seen by first combining any two of the three forces and then applying the special case (a) above. The vector diagram for the force summation will be a triangle, with the direction of the arrows following in order (see, for example, figure 2.9d).

Example 2.4

The motorcycle footbrake lever shown in figure 2.9a consists of a bellcrank pivoted to the chassis at O. If a vertical force of 75 N is applied by the rider's foot, what will be the tension T in the horizontal connecting cable? Find also the reaction on the motorcycle chassis at O.

In this example there are only three forces acting on the bellcrank, and these must be concurrent (see special case (b)), thus defining the line of action of R at O, as seen in figure 2.9b. At this stage the sense of R is unknown, and is therefore assumed, although with experience an intelligent guess may

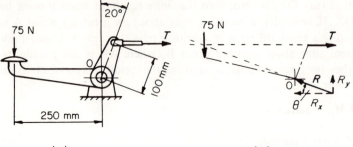

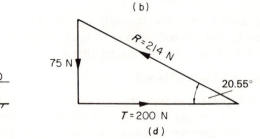

Figure 2.9

often be made. If the solution for R turns out to be positive, then the assumed sense was correct.

(a) *Analytical Solution*

$$\Sigma F_y = 0: \qquad R_y - 75 = 0 \qquad R_y = 75 \text{ N}$$

$$\Sigma F_x = 0: \qquad T - R_x = 0 \qquad R_x = T$$

$$\Sigma M_O = 0: \qquad 75 \times 250 - T \times 100 \times \cos 20° = 0$$

therefore

$$R_x = T = 200 \text{ N}$$

$$R = \sqrt{(R_x^2 + R_y^2)} = 214 \text{ N}$$

$$\theta = \tan^{-1}(R_y/R_x) = 20.55°$$

Note that the reaction R is that exerted by the frame on the bellcrank, which is equal and opposite to that on the chassis (Newton's third law), hence the required reaction is as shown in figure 2.9c.

(b) *Graphical Solution*

Having determined the line of the reaction R, a scaled force polygon may be drawn—a triangle in this case—see figure 2.9d. By measurement $T = 200$, $R = 214$ N and $\theta = 20.55°$, as before.

In this particular example there is little to choose between the two methods of solution, but this is not always the case. The best choice is largely a matter of experience.

Example 2.5

Movement of the bucket of the loader shown to scale in figure 2.10a is controlled by two identical linkages in parallel, only one of which is shown. The linkage is attached to the chassis at points F, J and K. The total weight of the bucket and its load is 5000 N and its centre of gravity is at G. For the position shown, determine the force in each of the three hydraulic cylinders BD, CF and HJ, and the reaction on the link KHE at K. The weights of all other parts are negligible.

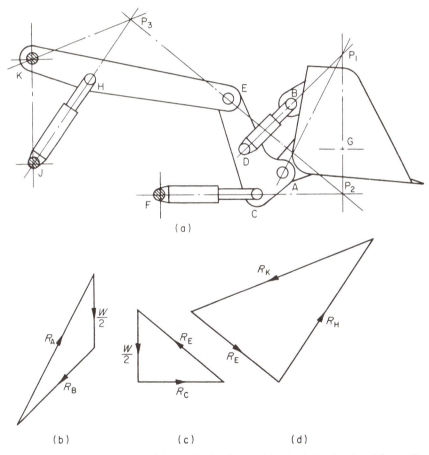

Figure 2.10   (b) Triangle of forces for bucket and load; (c) triangle of forces for bucket, load and link EDCA; (d) triangle of forces for link KHE

Consider first the bucket and load. This is subjected to three forces only, one at B, one at A, and its own weight, and these forces will therefore be concurrent. The line of action of the weight is given, and that of the force at B must be along the cylinder DB (DB is subjected to two forces only, one at each end, and these must be *collinear* by special case (a)). The intersection of these two lines of action defines the point P_1, and hence the line of action of the third force on the bucket, at A, must also pass through P_1. Knowing the magnitude of one of these forces, the weight (only half the weight has been actually used since there are two linkages), a triangle of forces can be drawn as shown in figure 2.10b.

Turning next to the assembly comprising bucket and load, link EDCA and cylinder DB, it is seen that this is also subjected to three forces only, namely the vertical load, a force at C and one at E. The lines of action of the first two intersect at P_2, hence giving the line of action of the third. Figure 2.10c shows the corresponding triangle of forces.

Finally consider the member KHE. The force at E is now known, the force at H will act along the line JH, defining point P_3, and hence the line of action of the remaining force, at K, can be drawn. The triangle of forces for KHE is shown in figure 2.10d.

These three force-triangles contain all the required information, which is easily scaled off, and is shown in figure 2.11. Care must be taken to ensure that the correct one of each of the pairs of equal and opposite forces at each connection is shown.

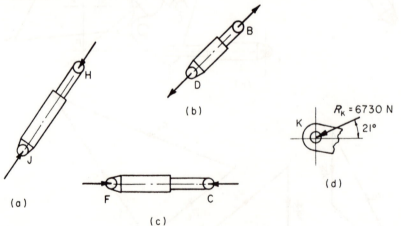

Figure 2.11   (a) Force in JH = 5770 N compressive; (b) force in DB = 3850 N tensile; (c) force in FC = 2880 N compressive; (d) force at K

Example 2.6

Figure 2.12a shows a uniform 250 kg platform supported by parallel links

AB and CD, each 0.8 m long. What torque, or moment, M, should be applied to CD to hold the system in equilibrium in the position shown?

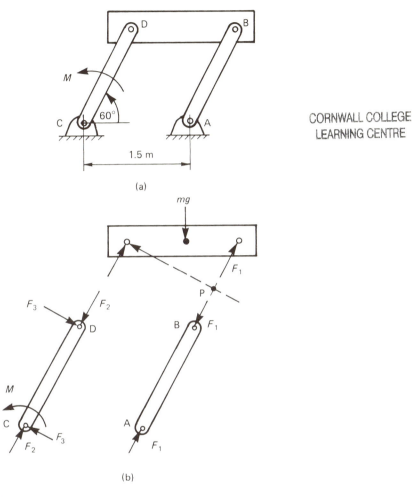

(a)

(b)

Figure 2.12

The member AB is in equilibrium under the action of two forces only, F_1 at each end, and these must be opposite and collinear. The same *cannot* be said for CD since, in addition to the forces at C and D, there is the moment M. It is convenient to represent the forces at C and D as components along and perpendicular to CD. These various forces are shown on the free-body diagrams, figure 2.12b.

Referring to the free-body diagram for the platform, and taking moments about the point P, anticlockwise positive

$$mg(0.75 - 1.5\cos 60°) - F_2(1.5\cos 30°) = 0$$

giving

$$F_2 = 655.4 \text{ N}$$

Now taking moments about B

$$mg(0.75) - 655.4(1.5\cos 30°) - F_3(1.5\cos 60°) = 0$$

hence

$$F_3 = 1317 \text{ N}$$

Finally, taking moments about C for the member CD

$$M - F_3(0.8) = 0 \qquad \text{hence} \qquad M = 1054 \text{ Nm}$$

Note that the solution has been obtained by moment equations only, two for the platform and one for the link CD. This was made possible by careful choice of the directions for the components of the force at D. Any other directions, say vertical and horizontal, would be equally valid, but might not result in quite such a simple solution. Careful attention to detail of this kind is frequently worthwhile.

2.3.2   Equilibrium in Three Dimensions

Equations 2.3 and 2.5 given in section 2.3.1 for equilibrium in two dimensions, in fact apply equally to three, although the graphical technique is generally less easy to use and visualise. In scalar terms, equilibrium in three dimensions requires the force summations in three directions, and the moment summations about three axes all to be zero. These directions and axes are normally chosen to be mutually perpendicular, giving, for example

$$\Sigma F_x = \Sigma F_y = \Sigma F_z = 0 \tag{2.6}$$

$$\Sigma M_x = \Sigma M_y = \Sigma M_z = 0 \tag{2.7}$$

There are thus now six independent equations or conditions of equilibrium to be satisfied.

Powerful vector techniques are available for the solution of three-dimensional problems, but they are beyond the scope of this text. Relatively simple problems may be solved analytically using equations 2.6 and 2.7, by considering orthogonal views of the problem. Graphical techniques may be useful in certain special cases, for example, balancing of rotating masses (which is covered in detail in chapter 12), where the forces act in parallel planes and all pass through points on a common axis, thus one force and one moment summation are necessarily zero, reducing the equilibrium conditions to four.

Example 2.7

Figures 2.13a and b show two views of a hand winch. In the position shown, what force P should be applied to raise a load of 35 kg? What will be the components of the forces on the shaft at the bearings A and B?

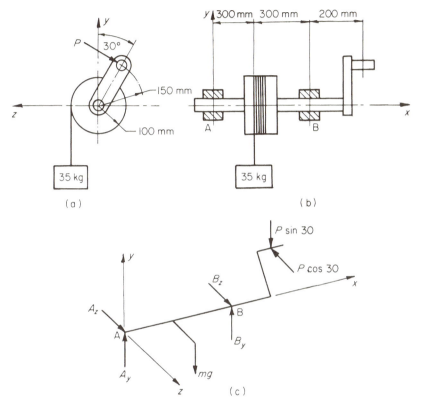

Figure 2.13

Figure 2.13c shows a free-body diagram for the shaft, drum and handle assembly.

$$\Sigma F_y: A_y + B_y - mg - P \sin 30° = 0$$

$$\Sigma F_z: A_z + B_z - P \cos 30° = 0$$

(there are no forces in the x-direction)

$$\Sigma M_x: 100mg - 150P = 0$$

$$\Sigma M_y: 600B_z - 800P \cos 30° = 0$$

$$\Sigma M_z: 300mg - 600B_y + 800P \sin 30° = 0$$

These solve to give

$$P = 229 \text{ N}$$

$$A_y = 134 \text{ N} \qquad A_z = -65 \text{ N}$$

$$B_y = 324 \text{ N} \qquad B_z = 264 \text{ N}$$

The minus sign for A_z indicates that the force is in the opposite direction to that shown in figure 2.13c.

Example 2.8

A large fire door weighing 15 000 N is hinged at two points A and B as shown in figure 2.14. Only the lower hinge B is capable of supporting any vertical load. Determine the components of the reactions on the door at A and B.

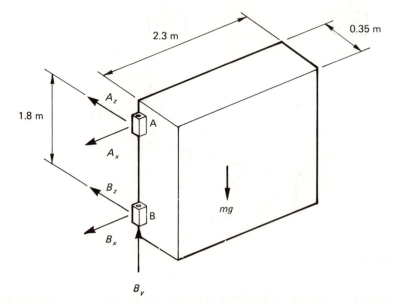

Figure 2.14

Summing the forces to zero

$$\Sigma F_x: A_x + B_x = 0 \qquad \Sigma F_y: B_y - mg = 0 \qquad \Sigma F_z: A_x + B_x = 0$$

and taking moments about axes through B

$$\Sigma M_x: 0.35mg/2 - 1.8A_z = 0 \qquad \Sigma M_z: 2.3mg/2 - 1.8A_x = 0$$

gives

$$A_x = -B_x = 9583 \text{ N} \qquad A_z = -B_z = 1458 \text{ N} \qquad B_y = 15\,000 \text{ N}$$

2.4 THE NATURE OF FRICTION

When there is a tendency towards relative motion between two bodies in contact, the reaction between them will have a component acting along the common tangent at the point of contact. This component is commonly referred to as a *friction force*, and is always in the direction that opposes the tendency to move. In some applications such as tyres, brakes and belt drives, this phenomenon is exploited, while in others, such as bearings and gears, it must be minimised. In yet other instances a balance must be found, for example with a screw thread there must be sufficient friction for the nut to remain tightened on the bolt, yet not so much that it is difficult to do up. Some of these particular applications are covered in later chapters.

Friction can be conveniently divided into several types.

(a) *Solid friction* (sometimes called *Coulomb friction* after an early experimenter in the field) exists between surfaces in rubbing contact.
(b) *Rolling friction* occurs where one member rolls without slip over another.
(c) *Fluid friction* arises where layers of fluid move over each other at different speeds. This last type is mentioned here only for completeness, but is discussed further in chapter 8 in the context of film lubrication.

2.4.1 Solid Friction

Consider the very simple experiment illustrated in figure 2.15a, such as that carried out by Coulomb and others. The horizontal force P applied to the block is slowly increased from zero until it starts to slide. The reaction R, from the surface onto the block, has been resolved into components, N vertically and F horizontally, the latter opposed to the ultimate direction of sliding, figure 2.15b. If the values of P and F are measured and plotted against one another, a curve as shown in figure 2.15c will be obtained. For low values of P the block will not move and F will be equal to P, that is, the block will be in equilibrium. At some particular value of P, the block will start to slide

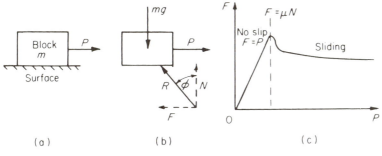

Figure 2.15 *Friction*

and continue to slide with increasing speed as P is increased still further, but generally this is accompanied by a *reduction* in F. This reduction is quite sharp at first, but then tends to level off.

From experiments such as these, some important results are established. For two particular surfaces in contact and on the point of slipping

(a) the friction force is independent of the apparent contact area
(b) the friction force is proportional to the normal force between the two surfaces.

This condition is normally referred to as *limiting friction*. The constant of proportionality is called the *coefficient of friction* and given the symbol μ, so that when about to slip

$$F = \mu N \tag{2.8}$$

From figure 2.15b

$$F/N = \tan \phi \tag{2.9}$$

where ϕ is known as the *friction angle*. Thus from equations 2.8 and 2.9, the value of ϕ is given by

$$\phi = \tan^{-1} \mu \tag{2.10}$$

The precise mechanism of friction is still not completely understood. There is some evidence to suggest that molecular attraction plays an important part, and that the independence of the contact area is only apparent. If the contact surfaces were greatly magnified, one would see that contact really only occurred over relatively few small areas, giving rise to local interference or even possibly small pressure welds. It has been suggested that it is the breaking of these bonds that gives rise to the phenomenon of friction. The presence of dirt, oxides, grease and moisture films are also relevant and can give rise to quite large fluctuations in the value of μ. Largely for this reason, experimental values are seldom quoted with any degree of accuracy.

Although previous discussion suggested that friction generally falls off with higher speeds (figure 2.15c) this effect is small and is seldom taken into account. Some examples of typical values of the coefficient of friction for common pairs of engineering materials are given in table 2.2.

In practice the coefficient of friction is usually less than unity, but there is no fundamental reason why this should be so.

Table 2.2   Coefficients of Friction

Materials	Dry	Coefficient of Friction Wet	Greasy
Metal on metal	0.15–0.3	0.3	0.05–0.08
Nylon on metal	0.1		
PTFE on metal	0.08		
Rubber on			
rough concrete	0.9–1.1	0.5–0.7	
Rubber on ice	0.1	0.02	
Asbestos brake or			
clutch lining on			
cast iron	0.2–0.4		

Example 2.9

A rectangular packing crate of mass m is placed on an incline of angle θ. What is the minimum necessary coefficient of friction if the crate is not to slide down the incline under its own weight? If the coefficient of friction is twice this value, what force must be applied horizontally to the crate to push it up the incline? Are there any circumstances in which it will not be possible to push the crate up the incline by the application of such a horizontal force?

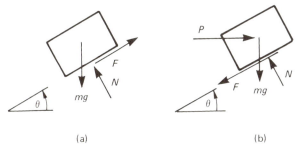

(a)   (b)

Figure 2.16

Figure 2.16a shows the free-body diagram for the initial solution. Since the crate may slide down the incline, the friction force must be up.
Summing forces parallel and perpendicular to the incline

$$F - mg \sin \theta = 0 \quad \text{and} \quad N - mg \cos \theta = 0$$

from which

$$F/N = \tan \theta$$

hence the minimum value of the coefficient of friction to prevent sliding is $\mu = \tan \theta$.

In the second part of the problem an attempt is being made to push the crate up the incline, so that now the friction force will be downwards, figure 2.16b, and the coefficient of friction is now $2 \tan \theta$. The force summations now give

$$P \cos \theta - F - mg \sin \theta = 0 \qquad \text{and} \qquad N - P \sin \theta - mg \cos \theta = 0$$

If the crate slips then

$$\mu = \frac{F}{N} = \frac{P \cos \theta - mg \sin \theta}{P \sin \theta + mg \cos \theta}$$

from which the required force P is defined by

$$\frac{P}{W} = \frac{\mu + \tan \theta}{1 - \mu \tan \theta} = \frac{3 \tan \theta}{1 - 2 \tan^2 \theta}$$

in this case. It is clear that if $2 \tan^2 \theta = 1$ then P will be infinite, which means that it will not be possible to push the crate up the incline no matter how large P is made. This phenomenon of lock-up is often exploited in ratchets and other devices.

Example 2.10

A motorcar of mass 980 kg has its mass centre 1.3 m behind the front axle, 1.5 m in front of the rear axle, and 0.8 m above ground level. The car is at the bottom of a slope of 20°, facing upwards. Assuming adequate engine power, investigate the possibility of driving up the hill (a) with rear-wheel drive and (b) with front-wheel drive, if the coefficient of friction between the driven wheels and the road is 0.8.

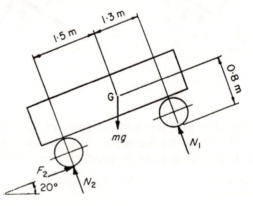

Figure 2.17

(a) Figure 2.17 shows a free-body diagram for the car attempting to drive up the hill with rear-wheel drive.

The direction of the friction force F_2 on the driven rear axle is determined as follows.

(i)   First imagine that slip does occur.
(ii) For the member on which the friction force is required (the tyre in this case), deduce the slipping direction relative to the other surface (the road).
(iii) Since this relative slipping direction is downhill, even though motion is hopefully uphill, the friction force on the tyre will be uphill as shown. There will, of course, be an equal and opposite friction reaction on the road, and it is this force that will cause any loose gravel to be ejected rearwards.

Summing forces parallel to the slope gives the required effort at the driving wheel, that is

$$F_2 - mg \sin 20° = 0$$

therefore

$$F_2 = 980 \times 9.81 \sin 20° = 3288 \text{ N}$$

Taking moments about the point of contact between the rear wheel and the road

$$1.5 \, mg \cos 20° - (1.5 + 1.3)N_1 - 0.8 \, mg \sin 20° = 0$$

Substituting numerical values gives

$$N_1 = 3900 \text{ N}$$

Summing forces perpendicular to the plane

$$mg \cos 20° - N_1 - N_2 = 0$$

therefore

$$N_2 = 5134 \text{ N}$$

Now the maximum friction force that can be sustained at the rear wheels is

$$\mu N_2 = 0.8 \times 5134 = 4107 \text{ N}$$

and since

$$F_2 < \mu N_2$$

drive is possible.
(b) In the second case, the force required up the plane is still 3288 N but now it must be generated at the front wheels. The maximum permissible friction force at the front is

$$\mu N_1 = 0.8 \times 3900 = 3120 \text{ N}$$

and hence front-wheel drive up the hill is not possible.

2.4.2 Rolling Friction

It is common experience that it requires more effort to pull a garden roller over a soft lawn that over concrete, yet in neither case is there any sliding. This occurs because no body is truly rigid and hence rolling will cause continuous deformation of both surfaces giving rise to *internal friction*.

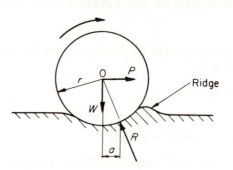

Figure 2.18   Rolling friction

Figure 2.18 shows this effect with the deformation of the surface under the rolling load greatly exaggerated. As the load W rolls to the right under the action of the force P, a ridge is pushed along in front giving a continuous resistance to motion. A single force R, equivalent to the distributed ground pressure will act typically as shown, and must pass through O. (Three forces in equilibrium are concurrent.) Taking moments about the point of application of R

$$Pr - Wa \approx 0$$

therefore

$$P \approx Wa/r$$

Now P must equal the horizontal component of R so that the ratio a/r is analogous to the coefficient of friction defined in the previous section. However, attempts to measure this quantity have not produced very consistent results, with values for metal on metal ranging from 10^{-3} to 10^{-6}, and for a pneumatic tyre on a metalled surface ranging from 0.01 to 0.05 at slow speeds.

SUMMARY

Newton's Laws

> Definition of equilibrium
> Units—in particular the kg, m, s system
> Actions and reactions

Weight

> $W = mg$

Centre of gravity and centre of mass

> $$x_G = \frac{\Sigma m_i x_i}{m} \qquad y_G = \frac{\Sigma m_i y_i}{m}$$

Free-body diagrams

Force summations

Moment of a force

Equilibrium

> $\Sigma F = 0 \qquad \Sigma M_O = 0$

Special cases of two and three forces in equilibrium

Solid friction

> In general $\quad F < \mu N$
> When slipping or about to slip $\quad F = \mu N$

Rolling friction

2.5   PROBLEMS

1. An astronaut has a mass of 72 kg. What is his weight on the earth? What is his 'weight' on the moon where the value of g is one-sixth that of the earth? When in earth orbit, astronauts are often described as 'weightless'; comment on the aptness of this term.
    [706.3 N, 117.7 N]

2. A road vehicle has two axles which are 3.4 m apart. The vehicle is positioned with its front axle on a weighbridge, which gives a reading of 12 850 N. When the rear axle is positioned on the weighbridge, the reading is 16 120 N. What is the longitudinal position of the centre of gravity of the vehicle?
    [1.89 m behind the front axle]

3. A 300 kg turbine rotor has its mass centre on its axis of rotation. Subsequently, a broken blade results in the loss of a mass of 195 g at a radius of 250 mm. How far off the axis of rotation will the centre of gravity now be?
   [0.163 mm]

4. Figure 2.19 shows a belt passing around two pulleys. If the tensions T_1 and T_2 are 450 and 260 N respectively, what are the torques on the pulleys and the reactions on the bearings at P and Q?
   [39.9 Nm, 14.25 Nm, 681.3 N]

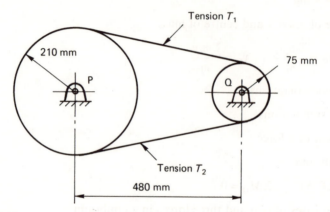

Figure 2.19

5. Four tugs are used to bring a liner to its berth as shown in figure 2.20. If each tug exerts a force of 5 kN along its own axis, determine the equivalent force–couple system at the foremast O, and the point on the hull where a single more powerful tug should push to produce the same effect.
   [13.34 kN, 46° to ship centre-line, 103.5 kNm, on the side of the hull 4.03 m aft of O]

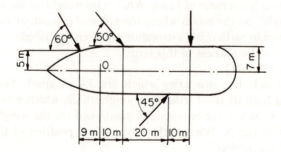

Figure 2.20

6. Find the components of the resultant of the system shown in figure 2.21.
[0, −63.3 N, −32.1 N; 75 Nm, 66 Nm, −32.5 Nm]

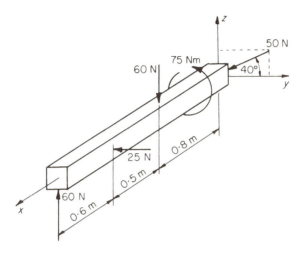

Figure 2.21

7. A thin uniform rod 240 mm long and having a mass of 1.5 kg is supported in a vertical plane as shown in figure 2.22. Assuming no friction find the reactions at A and B for each case.
[15.32 N, 4.25 N; 7.35 N, 7.35 N; 18.38 N, 6.36 N; 14.72 N, 0.882 Nm]

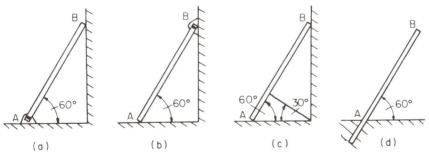

Figure 2.22

8. Figure 2.23 shows a 'magic tube' which defies all attempts to stand it up because of the overturning moment generated by the two hidden balls inside. What is the necessary ratio between the mass m of each ball and the mass M of the tube, in terms of the ball diameter d and the tube diameter D? Ignore any friction effects. How do you counter the argument that since the combined centre of gravity lies inside the supported area, it cannot fall over?
[$D/2[D − d]$]

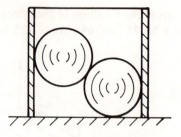

Figure 2.23

9. Figure 2.24 shows a vehicle suspension-unit. If the force on the tyre at the centre of the contact patch E has components of 5 kN vertically and 1 kN horizontally to the left, determine the forces on the body at A and B, and the force in the spring. The masses of the links and the wheel may be neglected, and the diagram may be scaled.
   [4.7 kN ∠17°, 5.5 kN 6°↗, 3.3 kN compressive]

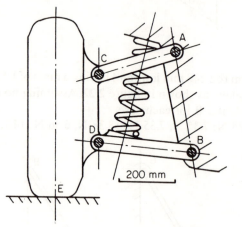

Figure 2.24

10. A uniform ladder 7 m long and having a mass of 40 kg leans at 30° to a vertical wall. The foot of the ladder rests on horizontal ground. If the coefficient of friction at the points where the ladder contacts the wall and the ground is 0.2, how far up the ladder may an 80 kg man climb before slip occurs?
    [2.15 m]

11. A simple hand-operated brake is shown in figure 2.25. What downward force P on the end of the layer is just sufficient to stop the wheel turning under the applied anticlockwise moment M? The coefficient of friction between the brake block and the wheel is μ. Show on the free-body diagram what happens if $\mu = a/c$. What effect does the direction of rotation have on the answers?
    [$M[a/\mu - c]/br$]

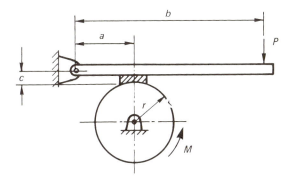

Figure 2.25

12. A uniform rod tests perpendicularly and symmetrically across two parallel rails, the coefficient of friction being the same at both points of contact. If an increasing horizontal force is applied at right angles to one end of the rod, show that slip will occur first at the rail nearer to the applied force.

13. Three identical storage drums are stacked as shown in figure 2.26. If the coefficient of friction between all pairs of contacting surfaces is the same, find the minimum necessary value if the stack is not to collapse. [*Hint*: note that the system may collapse because the upper drum *slides* down between the lower drums which *roll* apart, or because the upper drum *rolls* down between the lower drums which *slide* apart. While the problem may be solved by analysis of the various free-body diagrams, a very elegant solution can be obtained by careful study of its geometry.]
[$\tan 15° = 0.268$]

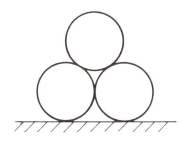

Figure 2.26

3 Kinematics

3.1 KINEMATICS OF A PARTICLE AND OF A LINE

Kinematics is concerned with the way in which bodies move and is essentially a branch of pure geometry. It deals with the relationships between displacement, velocity and acceleration, and the variation of these quantities with time. In this section, the study is confined to that of a particle and of a line.

3.1.1 Straight-line Motion of a Particle

Consider a particle P moving along a straight line, figure 3.1 so that at some time t its *displacement* from an arbitrary reference point O is x. If a small increase in displacement, δx, occurs in a time interval δt, then the *mean velocity* v_m of the particle during the interval is defined as

$$v_m = \delta x / \delta t$$

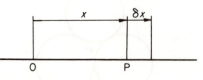

Figure 3.1   Straight-line motion

As the time interval is reduced, the mean velocity approaches the *velocity at an instant*, v, that is

$$v = \lim_{\delta t \to 0} \left(\frac{\delta x}{\delta t} \right) = \frac{dx}{dt} = \dot{x} \tag{3.1}$$

Suppose now that the velocity of the particle changes by an amount δv in the time interval δt, then the *mean acceleration*, a_m, is

$$a_m = \delta v / \delta t$$

and the *instantaneous acceleration, a*, is

$$a = \lim_{\delta t \to 0} \left(\frac{\delta v}{\delta t} \right) = \frac{dv}{dt} \tag{3.2}$$

Substituting for v from equation 3.1

$$a = d^2 x / dt^2 = \ddot{x} \tag{3.3}$$

and eliminating t from equations 3.1 and 3.2 gives

$$a = v \, dv / dx \tag{3.4}$$

The relations derived in equations 3.1 and 3.2 can be represented graphically in various ways, as in figure 3.2. Of these, the velocity–time graph will often be found most useful in providing a direct solution to a numerical problem. However, before equations 3.1 to 3.4 can be solved either analytically or graphically, some additional information is required.

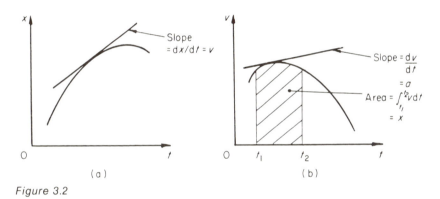

Figure 3.2

Example 3.1

A particle moves along a straight line such that after a time t seconds, its velocity v is

$$v = (10 - 6t) \text{ m/s}$$

What is the displacement of the particle after 4 seconds, measured from the point where t equals zero? Find also the total distance travelled in the 4 seconds.

Analytical Solution

From equation 3.1

$$v = dx / dt = 10 - 6t$$

therefore

$$\int_0^x dx = \int_0^4 (10 - 6t)\, dt$$

$$x = [10t - 3t^2]_0^4 = -8 \text{ m}$$

From inspection of the given data it is clear that the particle has an initial velocity of 10 m/s, which decreases to zero and then becomes negative. After 4 seconds the particle has returned to its starting point and moved a further 8 m in the opposite direction to which it started. At the end of its outward journey, the velocity of the particle is zero, so that

$$v = (10 - 6t) = 0$$

therefore the corresponding time is

$$t = 1.67 \text{ s}$$

and the displacement is

$$x = (10 \times 1.67) - (3 \times 167^2) = 8.33 \text{ m}$$

The total distance travelled is thus

$$d = 8.33 + 8.33 + 8 = 24.67 \text{ m}$$

Graphical Solution

First sketch the velocity–time relationship, showing the salient values, figure 3.3. The displacement after 4 seconds will be the net area, so that

$$x = \tfrac{1}{2} \times 10 \times \tfrac{5}{3} - \tfrac{1}{2} \times 14(4 - \tfrac{5}{3}) = -8 \text{ m}$$

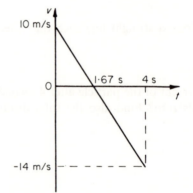

Figure 3.3

The distance travelled during 4 seconds will be the total area, so that

$$d = \tfrac{1}{2}(10 \times \tfrac{5}{3}) + \tfrac{1}{2} \times 14(4 - \tfrac{5}{3}) = 24.67 \text{ m}$$

(a) Motion Under Constant Acceleration

In the previous example, the acceleration is a constant. This can be seen by differentiating the expression for velocity or by noting that the velocity–time graph is a straight line. Such motion is, of course, a special case, but it arises so frequently that it is worth while developing general expressions relating the various quantities. Consider, therefore, a particle moving along a straight line with a constant acceleration a and having an initial velocity v_0 at zero time.

From equation 3.2

$$a = dv/dt$$

and since a is a constant in this case

$$\int_{v_0}^{v} dv = a \int_{0}^{t} dt$$

therefore

$$v = v_0 + a\, t \tag{3.5}$$

From equation 3.1

$$v = dx/dt$$

therefore

$$x = \int_{0}^{t} (v_0 + at)\, dt = v_0 t + \tfrac{1}{2}at^2 \tag{3.6}$$

From equation 3.4

$$a = v\, dv/dx$$

therefore

$$\int_{v_0}^{v} v\, dx = a \int_{0}^{x} dx$$

hence

$$\tfrac{1}{2}(v^2 - v_0^2) = ax$$

therefore

$$v^2 = v_0^2 + 2ax \tag{3.7}$$

Example 3.2

A city bus has a maximum speed of 90 km/h. It can accelerate from rest at a constant rate of 1.8 m/s², and can brake with a constant deceleration of 2.2 m/s² without discomfort to the passengers. How long does the bus take to cover a distance of 300 m, starting from and finishing at rest?

There are two possible situations depending on whether or not the bus has time to reach its maximum speed, as shown in figure 3.4. In figure 3.4b it is assumed that the bus will spend some time cruising at its maximum speed, while in figure 3.4a the distance is covered during acceleration and deceleration only.

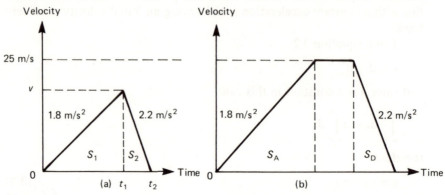

Figure 3.4

If the situation is correctly described by figure 3.4b then, using equation 3.7, the distances covered during acceleration and deceleration are given by

$$25^2 = 0 + 2 \times 1.8 \times s_A \qquad 0 = 25^2 - 2 \times 2.2 \times s_D$$

giving

$$s_A = 173.6 \text{ m} \qquad \text{and} \qquad s_D = 142.0 \text{ m}$$

Since the sum of these two distances exceeds the required 300 m, the situation must be described by figure 3.4a.

Using equation 3.5

$$v = 0 + 1.8t_1 \qquad \text{and} \qquad 0 = v - 2.2(t_2 - t_1)$$

Eliminating t_1 gives

$$v = (2.2 \times 1.8/4)t_2$$

Now the total distance travelled is the area under the v–t graph, so that

$$s = 300 \text{ m} = \tfrac{1}{2}vt_2$$

Now eliminating t_2

$$300 = \tfrac{1}{2}(2.2 \times 1.8/4)t_2^2$$

hence the total journey time is

$$t_2 = 24.6 \text{ s}$$

(b) Motion Under Variable Acceleration

In many problems the forces vary and therefore so also does the acceleration. A good example is given by a road vehicle where the driving force (due to engine torque) between the wheels and the road rises to a maximum and then decreases with increasing engine speed, while the resistance to motion will continually increase as some function of velocity. Solution of such problems must start from the basic equations 3.1 to 3.4.

Example 3.3

A vehicle accelerates from rest along a straight road, and its velocity is measured as it passes a series of posts spaced at ten-metre intervals. The data is tabulated below.

Post number	1	2	3	4	5	6	7	8	9	10
Velocity (km/h)	0	18	34	48	60	70	78	84	88	90

Estimate the time taken to travel from the third to the eighth post and the acceleration of the vehicle when it has travelled 55 m.

From equation 3.1

$$\int_0^t dt = \int_{x_1}^{x_2} (1/v)\,dx$$

where x_1 and x_2 are the displacements at the third and eighth posts, that is, 20 m and 70 m respectively. From the data, a graph of $1/v$ against x can be plotted, and the appropriate area estimated (figure 3.5a).

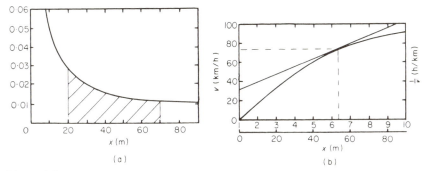

Figure 3.5

$$t = \int_{20}^{70} (1/v)\, dx = 3.06 \text{ s}$$

From equation 3.4

$$a = v\, dx/dx$$

Plotting the data as given, figure 3.5b, the slope, that is dv/dx at x equals 55 m is easily obtained so that

$$a = \frac{75}{3.6}\left(\frac{100-32}{85}\right)\frac{1}{3.6} = 4.63 \text{ m/s}^2$$

Example 3.4

Derive expressions for the displacement, velocity and acceleration of a particle which moves along the x-axis so that its acceleration is always proportional to its distance from the origin, and directed towards it. Assume that when $t = 0$, $x = 0$, and $v = v_0$.

Let the constant of proportionality be k so that

$$a = -kx$$

Substituting in equation 3.4

$$-kx = v\, dv/dx$$

so that, integrating

$$-\tfrac{1}{2}kx^2 = \tfrac{1}{2}v^2 + \text{constant}$$

When $x = 0$, $v = v_0$, giving

$$v = v_0 \sqrt{\left(1 - \frac{kx^2}{v_0^2}\right)}$$

$$= dx/dt$$

therefore

$$t = \frac{1}{v_0}\int_0^x \frac{1}{\sqrt{\left(1 - \dfrac{kx^2}{v_0^2}\right)}}\, dx$$

This is a standard integral whose solution gives

$$x = \frac{v_0}{\sqrt{k}} \sin \sqrt{k} \, t$$

Differentiating gives

$$v = v_0 \cos \sqrt{k} t$$

and differentiating again

$$a = -v_0 \sqrt{k} \sin \sqrt{k} \, t$$

Motion of this kind is commonly called *simple harmonic* and will be developed further in the study of vibrations in chapters 14–16.

Example 3.5

In the absence of air resistance, a freely falling body will have a constant downward acceleration g. In reality, the air resistance will increase with velocity, usually in a non-linear manner. Examine the motion of an object released from rest where the resistance varies with the square of the velocity, so that the acceleration is $g - kv^2$, where k is a constant.

By definition, when the acceleration is zero the velocity must be constant. Thus, when $g - kv^2 = 0$, $v = v_t = \sqrt{(g/k)}$. This is the *terminal velocity*, that is, the maximum velocity the object will achieve.

Using equation 3.2

$$a = \frac{dv}{dt} = g - kv^2$$

Separating the variables

$$\int_0^t dt = \int_0^v \frac{dv}{g - kv^2}$$

This is a standard integral that can be found in any appropriate level mathematics book. Alternatively, the integral can be evaluated from first principles by noting that

$$\frac{1}{1 - u^2} = \frac{1}{2(1 - u)} + \frac{1}{2(1 + u)}$$

Thus

$$t = \frac{v_t}{2g} \log_e \left[\frac{1 + v/v_t}{1 - v/v_t} \right]$$

which can be arranged to give

$$v = v_t \left[\frac{1 - e^{-2gt/v_t}}{1 + e^{-2gt/v_t}} \right]$$

Using equation 3.1

$$\int_0^x dx = \int_0^t v_t \left[\frac{1 - e^{-2gt/v_t}}{1 + e^{-2gt/v_t}} \right] dt$$

This can also be integrated by making the substitution $1 + e^{-2gt/v_t} = u$ and using partial fractions to give

$$x = \frac{v_t^2}{g} \left\{ \log_e \left[\frac{1 + e^{-2gt/v_t}}{2} \right] + gt/v_t \right\}$$

Figure 3.6 shows the form of the acceleration, velocity and displacement as functions of time.

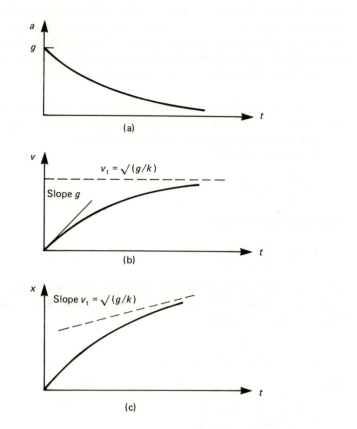

(a)

(b)

(c)

Figure 3.6

3.1.2 Angular Motion of a Line

Consider a line PQ whose *angular displacement* at some time *t* from an arbitrary reference axis is θ, as shown in figure 3.7. Now suppose that during

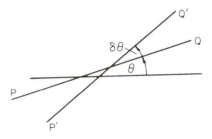

Figure 3.7 Angular motion of a line

a small increment of time δt, the line PQ moves through a small angle $\delta\theta$ to a new position P'Q'. Using a similar technique to that used in the previous section for straight-line motion, the *mean angular velocity* ω_m is

$$\omega_m = \delta\theta/\delta t$$

and the *instantaneous angular velocity* is

$$\omega = \lim_{\delta t \to 0} \left(\frac{\delta\theta}{\delta t} \right) = \frac{d\theta}{dt} = \dot{\theta}$$

Continuing the analogy, the *angular acceleration* α is

$$\alpha = d\omega/dt = d^2\theta/dt^2 = \ddot{\theta}$$

or

$$\alpha = \omega \, d\omega/d\theta$$

These results may be expressed graphically in a manner very similar to that of figure 3.2 replacing the linear quantity by the corresponding angular quantity.

It is essential that in angular motion problems radian units are used for all angular measures.

Example 3.6

For tests on an electric motor, the armature is accelerated uniformly from rest to a speed of 1450 rev/min, held at this speed for 10 s, and then decelerated uniformly to rest. If the test is to last for 800 revolutions, what is the total duration *t*?

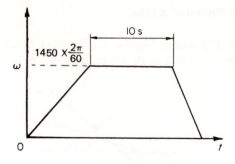

Figure 3.8

Figure 3.8 shows the shape of the angular velocity–time relationship. (Notice that with these axes the constant acceleration and deceleration give straight lines although their slopes cannot be determined.)

$$\theta = 800 \times 2\pi \text{ rad} = \int_0^t \omega \, dt = \text{area under } \omega - t \text{ graph}$$

Taking the area of a trapezium as height × mean length

$$800 \times 2\pi \text{ rad} = \frac{1}{2}\left(1450 \times \frac{2\pi}{60}\right)(t + 10)$$

therefore

$$t = 56.2 \text{ s}$$

Alternatively, the time could be found by using the angular equivalents of the relations determined in equations 3.5 to 3.7.

Example 3.7

The trap door shown in figure 3.9 is given a very small displacement to just start it moving. What will be its angular velocity just before it reaches the

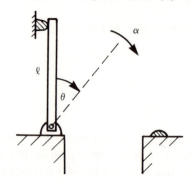

Figure 3.9

stop when $\theta = 90°$? The angular acceleration of the door due to its own weight is $\alpha = (3g/2l) \sin \theta$, where l is the length of the door and θ is its angular position measured from the vertical.

The angular equivalent of equation 3.4 gives

$$\alpha = (3g/2l) \sin \theta = \omega \, d\omega/d\theta$$

from which

$$\int_0^{\pi/2} (3g/2l) \sin \theta \, d\theta = \int_0^{\omega} \omega \, d\omega$$

Integrating and rearranging

$$\omega = \sqrt{(3g/l)}$$

3.1.3 Curvilinear Motion of a Particle

When a particle moves along a curved path in a plane, it will require two co-ordinates to define its position at any instant, and the way these co-ordinates change with time will determine the particle's velocity and acceleration. The choice of co-ordinate system in any particular case will depend on the nature of the problem and on the way the data is presented. Two relatively common co-ordinate systems will now be considered in some detail.

(a) *Cartesian Co-ordinates*

In figure 3.10 the particle P is following a curved path in the x–y plane, and its position at some instant is defined by the co-ordinates (x, y). It is clear

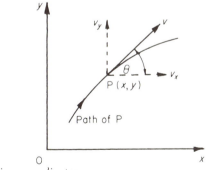

Figure 3.10 *Cartesian co-ordinates*

that the velocity v is always tangential to the path of P, and at the instant shown can be resolved into components

$$v_x = v \cos \theta$$

$$v_y = v \sin \theta$$

From equations 1.1 and 1.2

$$v = \sqrt{(v_x^2 + v_y^2)}$$

$$\theta = \tan^{-1} v_y/v_x$$

The results of section 3.1.1 may then be applied independently in the x- and y-directions, with time as the linking parameter.

Example 3.8

A gun is to be laid so that the shell will strike a target 3000 m away in the horizontal direction and 240 m higher in elevation. If the muzzle velocity is 850 m/s, determine the time t taken for the shell to reach the target, the barrel elevation θ, and the velocity of the shell v, when it strikes. Assume that air resistance is negligible so that the horizontal component of velocity is constant and the vertical acceleration is 9.81 m/s² downwards due to gravity. The results of section 3.1.1a are then applicable.

Referring to figure 3.11

$$3000 = 850 \cos \theta \, t$$

$$240 = 850 \sin \theta \, t - \tfrac{1}{2} \times 9.81 \, t^2$$

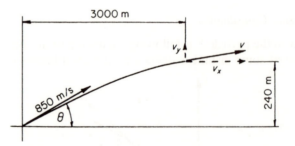

3000 m

240 m

850 m/s

θ

v_y

v_x

v

Figure 3.11

Eliminating θ gives

$$t^4 - 4t^2(850^2 - 240 \times 9.81)/9.81^2 + 4(3000^2 + 240^2)/9.81^2 = 0$$

which when solved gives

$$t = 173 \text{ s} \qquad \text{and} \qquad 3.55 \text{ s}$$

The first solution corresponds to a very high trajectory with the barrel almost vertical, and is rather impractical. The second solution is the more likely and can be substituted back to give

$$\cos \theta = 3000/(850 \times 3.55)$$

therefore

$$\theta = 6.17°$$

Since the horizontal velocity component is constant

$$v_x = 850 \cos 6.17° = 845 \ \text{m/s}$$

Also, from

$$v_y^2 = 850^2 \sin^2 6.17° - 2 \times 9.81 \times 240 = 3637 \ \text{m}^2/\text{s}^2$$

$$v_y = 60.3 \ \text{m/s}$$

and

$$v = \sqrt{(845^2 + 60.3^2)} = 847 \ \text{m/s}$$

Example 3.8 illustrates an important class of problem commonly described as *projectile motion*, and merits further study. In particular, the nature of the trajectory, the maximum altitude and the range are of interest, figure 3.12.

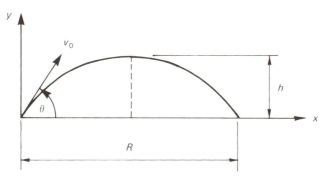

Figure 3.12

Applying the results of section 3.1.1a

$$x = v_{0x}t \qquad \text{and} \qquad y = v_{0y}t - \tfrac{1}{2}gt^2$$

where v_0 is the initial velocity tangential to the trajectory, and the suffixes x and y refer to the horizontal and vertical directions respectively. Note that upwards is positive so that the vertical acceleration is $-g$.

Eliminating t from the expressions for x and y gives

$$y = x \tan \theta - x^2 \frac{g}{2v_0^2 \cos^2 \theta} \tag{3.8}$$

where θ is the angle between the initial velocity and the horizontal. It is clear from the above expression that the trajectory is parabolic.

The horizontal velocity component is

$$v_y = v_{0y} - gt$$

and at the highest point of the trajectory this will be zero, thus

$$t_h = v_{0y}/g$$

where t_h is the time to reach the maximum altitude h.

Substituting this particular value of time into the general expression for y and rearranging gives

$$h = \frac{v_{0y}^2}{2g} = \frac{v_0^2 \sin^2 \theta}{2g} \qquad (3.9)$$

The maximum range will occur when $y = 0$. Substituting this condition into equation 3.8 and rearranging gives

$$R = \frac{v_0^2 \sin(2\theta)}{g} \qquad (3.10)$$

It must be emphasised that this treatment of projectiles is for motion under gravity alone, that is, the effect of air resistance is neglected.

(b) Polar Co-ordinates

Figure 3.13a shows a particle P moving along a curve with its position at some time t defined by the radius vector r measured from some convenient datum O. Axes have been set up at P along the radius (radial or r-axis), and perpendicular to the radius (transverse or θ-axis). It is clear that in general the motion of P will have components in both the radial and transverse directions.

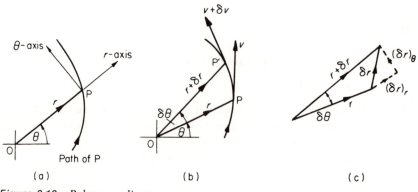

Figure 3.13   Polar co-ordinates

Consider the particle moving around the curve from P to P′ in a time interval δt while its position and velocity change as shown in figure 3.13b.

By taking the vector difference of the position vectors, the displacement δr can be obtained, figure 3.13c, and this can be resolved into the r- and θ-directions to give $(\delta r)_r$ and $(\delta r)_\theta$ respectively.

Now

$$(\delta r)_r \approx \delta r$$

and

$$(\delta r)_\theta \approx r\,\delta\theta$$

so that in the limit, the velocity components become

$$v_r = \mathrm{d}r/\mathrm{d}t = \dot{r} \qquad (3.11)$$

and

$$v_\theta = r\,\mathrm{d}\theta/\mathrm{d}t = r\dot{\theta} \qquad (3.12)$$

To find the acceleration components, consider first the change in the radial velocity as shown in figure 3.14a. This change, δv_r, has components in both the radial and transverse directions, namely

$$(\delta v_r)_r \approx \delta v_r$$

and

$$(\delta v_r)_\theta \approx v_r\delta\theta$$

and therefore due to these changes, there will be acceleration terms

$$a_r = \mathrm{d}v_r/\mathrm{d}t = \mathrm{d}^2r/\mathrm{d}t^2 = \ddot{r} \qquad (3.13)$$

and

$$a_\theta = v_r\,\mathrm{d}\theta/\mathrm{d}t = \dot{r}\dot{\theta} \qquad (3.14)$$

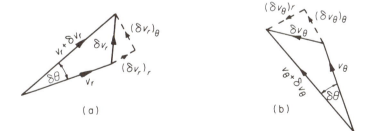

Figure 3.14 *Polar co-ordinates*

Next consider the change in transverse velocity, figure 3.14b. This change δv_θ also has components in both the radial and transverse directions

$$(\delta v_\theta)_r \approx -v_\theta \delta\theta$$

and

$$(\delta v_\theta)_\theta \approx \delta v_\theta$$

(the minus sign for $(\delta v_\theta)_r$ arising because it is in the opposite direction to r). Using equation 3.12, these two velocity changes give two more acceleration terms

$$a_r = -v_\theta \, d\theta/dt = -r\dot{\theta}^2 \tag{3.15}$$

and

$$a_\theta = dv_\theta/dt = d(r\dot{\theta})/dt = r\ddot{\theta} + \ddot{r}\dot{\theta} \tag{3.16}$$

Combining equations 3.13 and 3.15, and equations 3.14 and 3.16, the final acceleration components

$$a_r = \ddot{r} - r\dot{\theta}^2 \tag{3.17}$$

and

$$a_\theta = r\ddot{\theta} + 2\dot{r}\dot{\theta} \tag{3.18}$$

are obtained. The term $2\dot{r}\dot{\theta}$ is commonly called the *Coriolis component of acceleration*.

The velocity and acceleration components in polar co-ordinates are summarised in figure 3.15.

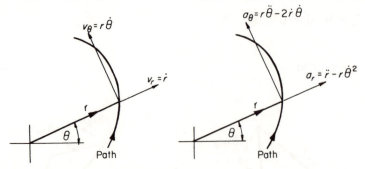

Figure 3.15   Velocity and acceleration components in polar co-ordinates

Example 3.9

The lawn sprinkler shown in figure 3.16a has two horizontal radial arms each 150 mm long that rotate about a vertical axis. Water flows out along the arms at a constant radial velocity of 0.8 m/s and the arms rotate at a constant angular velocity of 300 rev/min. What is the acceleration of an element of water (a) just before it leaves the end of the arm, and (b) just after it leaves the end of the arm?

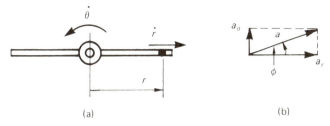

(a)   (b)

Figure 3.16

(a) From equations 3.17 and 3.18

$$a_r = 0 - r\dot{\theta}^2 = -0.150 \times (300 \times 2\pi/60)^2 = 148.0 \text{ m/s}^2$$

$$a_\theta = 0 + 2\dot{r}\dot{\theta} = 2 \times 0.8 \times (300 \times 2\pi/60) = 50.3 \text{ m/s}^2$$

Hence the total acceleration just before leaving the end of the arm (figure 3.16b) is

$$a = \sqrt{(148.0^2 + 50.3^2)} = 156.3 \text{ m/s}^2$$

$$\phi = \tan^{-1}(a_\theta/a_r) = 18.8°$$

(b) As soon as the water leaves the end of the arm, it is subjected to gravity only, and hence its acceleration will be g downwards.

A special case of curvilinear motion is that of a particle moving in a circle. Since r is then a constant, both the radial velocity $\dot{r}$ and the radial acceleration $\ddot{r}$ are zero, and thus equations 3.11, 3.12, 3.17 and 3.18 reduce to

$$v_r = 0 \quad \text{and} \quad v_\theta = v = r\dot{\theta} \qquad\qquad (3.19)$$

$$a_r = -r\dot{\theta}^2 = -v^2/r \quad \text{and} \quad a_\theta = r\ddot{\theta} \qquad\qquad (3.20)$$

as shown in figure 3.17. Note that the negative sign for a_r means that the radial component of acceleration of a particle moving in a circle is directed towards the centre of rotation, and is commonly referred to as a *centripetal acceleration*.

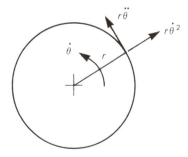

Figure 3.17 Circular motion

Example 3.10

A pilot flies his aircraft in a vertical loop of 450 m radius. At the top of the loop he notes that he is just 'weightless' (see section 2.1.3). What is the speed of the aircraft?

The sensation of weightlessness arises because the pilot is in free fall, that is, his acceleration is g downwards, and since this occurs at the top of the loop, it is also his radial acceleration.

Thus

$$g = v^2/r$$

so that

$$v = \sqrt{(9.81 \times 450)} = 66.4 \text{ m/s} = 240 \text{ km/h}$$

3.2 KINEMATICS OF A RIGID BODY

The motion of a rigid body in a plane can be uniquely defined by the motion of some convenient point together with the angular motion of the body. In practice the point of interest is usually the mass centre, but whichever point is chosen the techniques developed for a particle in the previous sections are relevant. Also, since the angular displacement between two lines drawn on the body does not change, the angular motion can be defined by considering *any* line on the body. Before looking at particular examples it will be helpful to classify the motion.

(a) *Translation* occurs when all particles move in parallel paths, for example the piston in an IC engine (rectilinear translation), or the connecting link of a railway locomotive (curvilinear translation). In these cases *every point has the same motion* and thus *any point* may be used to define the motion of the body as a whole.

(b) *Fixed-axis rotation* occurs when all particles move in concentric circular paths such as a motor armature or the turbine rotor of a stationary powerplant. In such cases only the angular motion need be defined.

(c) *General plane motion* is a combination of translation and rotation, and the motions can be considered quite separately. Examples are a rolling wheel and the connecting-rod of an IC engine.

Example 3.11

A wheel of radius R rolls without slip along a straight path with an angular velocity ω and angular acceleration α. Derive an expression for the velocity and acceleration of the geometric centre, and for the velocity of a general point on the rim. Which point on the rim has zero velocity?

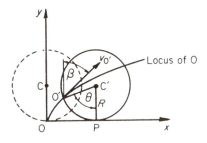

Figure 3.18

Referring to figure 3.18, the wheel has rolled from the initial position shown dotted to that shown in full, the centre C moving to C′ and point O moving to O′, where the line O′C′ makes an angle θ with the vertical. The $x-y$ co-ordinates of the centre C′ are

$$x_{C'} = OP = \text{arc } PO' = R\theta$$

$$y_{C'} = C'P = R$$

Differentiating

$$\dot{x}_{C'} = R\dot{\theta} = R\omega$$

$$\dot{y}_{C'} = 0$$

and

$$\ddot{x}_{C'} = R\ddot{\theta} = R\alpha$$

$$\ddot{y}_{C'} = 0$$

In a similar manner the $x-y$ co-ordinates of the general point O′ are

$$x_{O'} = OP - O'C' \sin\theta = x_{C'} - R\sin\theta$$

$$y_{O'} = C'P - O'C' \cos\theta = R(1 - \cos\theta)$$

Differentiating

$$\dot{x}_{O'} = \dot{x}_{C'} - R\cos\theta\,\dot{\theta} = R\omega(1 - \cos\theta)$$

$$\dot{y}_{O'} = R\sin\theta\,\dot{\theta} = R\omega\sin\theta$$

hence

$$v_{O'} = R\omega\sqrt{[(1 - \cos\theta)^2 + \sin^2\theta]} = R\omega\sqrt{[2(1 - \cos\theta)]}$$

and the angle between $v_{O'}$ and the vertical is

$$\beta = \tan^{-1}\dot{x}_{O'}/\dot{y}_{O'} = \tan^{-1}[(1 - \cos\theta)/\sin\theta] = \theta/2$$

It is clear that for $v_{O'} = 0$, $\cos\theta = 1$ and $\theta = 0$ hence point P, where the wheel contacts the ground, has zero velocity. Such a point is known as the *instantaneous centre of zero velocity* and will be considered in more detail in section 3.3.1.

Example 3.12

In figure 3.19 the lower end B of the uniform rod AB moves out along the *x*-axis with a velocity of 2 m/s and an acceleration of 0.5 m/s² while end A is constrained to move along the *y*-axis. For the position where θ is 60°, find the angular acceleration of the rod and the components of the acceleration of the mass centre G.

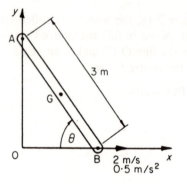

Figure 3.19

For the end B in a general position defined by the angle θ

$$x_B = 3 \cos \theta$$

hence

$$\dot{x}_B = -3 \sin \theta \, \dot{\theta} = 2 \text{ m/s}$$
$$\ddot{x}_B = -3 \sin \theta \, \ddot{\theta} - 3 \cos \theta \, \dot{\theta}^2 = 0.5 \text{ m/s}^2$$

therefore

$$\dot{\theta} = -0.770 \text{ rad/s}$$
$$\ddot{\theta} = -0.535 \text{ rad/s}^2$$

For the mass centre G

$$x_G = 1.5 \cos \theta$$
$$y_G = 1.5 \sin \theta$$

hence

$$\dot{x}_G = -1.5 \sin \theta \, \dot{\theta}$$
$$\dot{y}_G = 1.5 \cos \theta \, \dot{\theta}$$

and

$$\ddot{x}_G = -1.5 \sin \theta \, \ddot{\theta} - 1.5 \cos \theta \, \dot{\theta}^2$$
$$\ddot{y}_G = 1.5 \cos \theta \, \ddot{\theta} - 1.5 \sin \theta \, \dot{\theta}^2$$

from which

$$\ddot{x}_G = 0.25 \text{ m/s}^2$$
$$\ddot{y}_G = -1.03 \text{ m/s}^2$$

3.3 RELATIVE MOTION

In the preceding sections on kinematics, displacements velocities and accelerations have all been absolute, that is, they have been measured from a stationary reference. It is often useful to consider the motion of one point *relative* to another point that has a different motion.

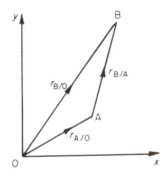

Figure 3.20 *Relative motion*

Consider two points A and B whose displacements from a fixed origin O are $r_{A/O}$ and $r_{B/O}$ respectively (figure 3.20). Then the displacement of B relative to A is the vector difference, that is

$$r_{B/A} = r_{B/O} - r_{A/O}$$

therefore

$$r_{B/O} = r_{B/A} + r_{A/O}$$

Differentiating

$$v_{B/O} = v_{B/A} + v_{A/O} \tag{3.21}$$

and

$$a_{B/O} = a_{B/A} + a_{A/O} \tag{3.22}$$

Since O is a fixed point, the subscripts B/O and A/O indicate absolute quantities and are usually reduced to B and A respectively. Also it is clear that the process can be extended indefinitely so that in general

$$r_Q = r_{Q/A} + r_{A/B} + r_{B/C} + \ldots + r_P$$

$$v_Q = v_{Q/A} + v_{A/B} + v_{B/C} + \ldots + v_P$$

$$a_Q = a_{Q/A} + a_{A/B} + a_{B/C} + \ldots + a_P$$

Example 3.13

Two winches A and B, each of 250 mm radius, are used to hoist a steel bar by means of cables attached at C and D, figure 3.21. At the instant shown winch A starts to rotate anticlockwise with an angular acceleration of 0.4 rad/s² while winch B starts to rotate clockwise with an angular acceleration of 0.6 rad/s². Determine the angular acceleration of the bar and the acceleration of its mass centre G.

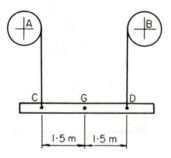

Figure 3.21

Considering a point on the periphery of each winch and using equation 3.20

$$a_C = r\ddot{\theta}_C = 0.25 \times 0.4 = 0.1 \text{ m/s}^2$$

$$a_D = r\ddot{\theta}_D = 0.25 \times 0.6 = 0.15 \text{ m/s}^2$$

where vertically upwards is taken as positive. Now

$$a_{D/C} = a_D - a_C = 0.05 \text{ m/s}^2$$

and considering this relative motion in $r - \theta$ co-ordinates (section 3.1.3b)

$$a_{D/C} = DC\ddot{\theta}$$

hence

$$\ddot{\theta} = 0.05/3 = 0.017 \text{ rad/s}^2$$

and by inspection $\dot{\theta}$ is anticlockwise. From equation 3.22

$$a_G = a_{G/C} + a_C$$
$$= GC\dot{\theta} + a_C = 1.5 \times 0.017 + 0.1$$
$$= 0.125 \text{ m/s}^2$$

3.3.1 Instantaneous Centre

In example 3.11 it was seen that the point at the bottom of the wheel had no velocity. This could have been deduced directly since any two points in contact without slip must have a common velocity, and clearly one of these points, that on the horizontal surface, is at rest.

Equally clearly, the point on the rolling wheel which is in contact with the ground and has no velocity is constantly changing. Notwithstanding this, if such an instantaneous centre of zero velocity can be found on a body, then, *for velocity calculations*, the body can be treated as if it were rotating about a fixed point coincident with the instantaneous centre, *but at the instant shown only*. (An instantaneous centre of zero acceleration can also be defined, but it will not generally coincide with the instantaneous centre of zero velocity, and its use is not common. In this text, instantaneous centre will be assumed to apply to velocity only.)

Example 3.14

Rework the velocity part of example 3.11 using the instantaneous centre concept.

Referring to figure 3.22, P is the instantaneous centre, and since all lines on a rigid body have the same angular motion, equation 3.19 can be used directly to give

$$v_{C'} = \dot{x}_{C'} = R\omega$$

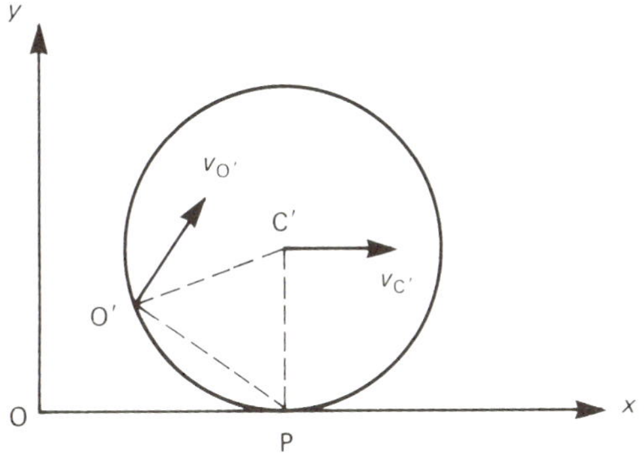

Figure 3.22

and is perpendicular to PC' as shown. In a similar manner

$$v_{O'} = O'P\omega = 2R\omega \sin(\theta/2) = R\omega\sqrt{[2(1-\cos\theta)]}$$

and is perpendicular to $O'P$.

These answers are identical to those obtained previously, and it is clear that, if the instantaneous centre can be found, the subsequent solution is often simpler.

If the velocity of one point on a body is known completely, and the line of the velocity of another point is known, then the instantaneous centre can always be found. Considering example 3.12 and referring to figure 3.23, it is

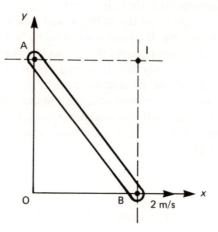

Figure 3.23

clear that at the instant shown, *point A is moving as if it were rotating about some point on a line through A perpendicular to the y-axis.* Point B can be considered in a similar manner, giving another line intersecting the first at the instantaneous centre I. Even though I lies outside the body, we can say that the angular velocity of AB is

$$\dot{\theta} = \dot{x}_{\mathrm{B}}/\mathrm{BI} = \tfrac{2}{3}\sin 60° = 0.770 \,\mathrm{rad/s}$$

and by inspection this is anticlockwise.

SUMMARY

Straight-line motion

  Always true

$$a = \frac{dv}{dt} = \frac{v\,dv}{dx} = \frac{d^2x}{dt^2} = \text{Area under velocity–time graph}$$

$$v = \frac{dx}{dt} = \text{Slope of displacement–time graph}$$

$x = $ Area under velocity–time graph

  Constant acceleration

$$v = v_0 + at \qquad v^2 = v_0 + 2ax \qquad x = v_0 t + \tfrac{1}{2}at^2$$

  Simple Harmonic Motion

$$a = -kv = -v_0\sqrt{k}\sin\sqrt{k}t$$

$$v = v_0(1 - kx^2/v_0) = v_0\cos\sqrt{k}t$$

$$x = (v_0/\sqrt{k})\sin\sqrt{k}t$$

  Angular motion of a line

    Replace linear quantities above by corresponding angular
    quantities
    Use radians

Curvilinear motion

  Cartesian coordinates

$$v_x = v\cos\theta \qquad v_y = v\sin\theta$$

$$v = \sqrt{(v_x^2 + v_y^2)} \qquad \theta = \tan^{-1}(v_y/v_x)$$

  Polar co-ordinates

$$a_r = \ddot{r} - r\dot{\theta}^2 \qquad a_\theta = r\ddot{\theta} + 2\dot{r}\dot{\theta}$$

  Special case of circular motion

$$v_r = 0 \qquad v_\theta = r\dot{\theta}$$

$$a_r = -r\dot{\theta}^2 = -v_\theta^2/r \qquad a_\theta = r\ddot{\theta}$$

Rigid body motion

    Translation
    Fixed axis rotation
    General plane motion

Relative motion

Instantaneous centre of zero velocity

3.4   PROBLEMS

1. A particle starts from rest and moves such that the product of its velocity and acceleration is always 5 m²/s³. How far will it travel in 10 s?
   [66.7 m]

2. The deceleration of a flywheel due to friction and wind resistance is proportional to the square of its angular velocity. If the flywheel slows down from 1000 rev/min to 800 rev/min in 12 s, how much longer will it take to slow down to 600 rev/min?
   [20 s]

3. A public-service vehicle has an acceleration of 4.2 m/s², a top speed of 15 m/s and a deceleration of 3.8 m/s². How long does it take to travel between two stops 600 m apart?
   [43.8 s]

4. A particle starts to move from rest along a straight line so that its acceleration varies with time as given below.

acceleration (m/s²)	0	1.74	3.83	.5.85	7.37	7.98	7.23	4.72	0	−7.36	−17.80
time (s)	0	1	2	3	4	5	6	7	8	9	10

   Find the maximum velocity, and the displacement after 10 seconds.
   [39.3 m/s, 207 m]

5. A motorcyclist accelerates uniformly from rest to a speed of 150 km/h and then immediately applies the brakes to bring the machine to a stop with a constant deceleration. If the total distance travelled is 700 m, determine the time taken.
   [33.6 s]

6. In the cathode-ray tube shown in figure 3.24, electrons enter the gap between the charged plates with a velocity v_0, while between the plates the force on each electron gives it a constant lateral acceleration a, and

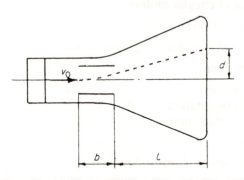

Figure 3.24

once clear the electrons travel in straight lines. Determine the deflection
d at the tube face.
$$[(ab/v_0^2)(l + b/2)]$$

7. An anti-tank shell has a muzzle velocity of 800 m/s and is to be fired at
a target which the gunner estimates to be 1200 m distant over level
ground. The barrel is elevated to an angle of 0.5° above the horizontal
which puts the muzzle 2.6 m above ground level. If the target is 3.5 m
high, to what accuracy should the range have been estimated to ensure
a hit? Neglect air resistance.
[Estimate must be between 1024 m and 1384 m]

8. A particle P moves along a line OQ which is rotating in a vertical plane
about a fixed axis O, figure 3.25. In the position shown, the absolute
velocity of the particle is 50 m/s horizontally to the right and its absolute
acceleration is 2000 m/s² vertically downwards. Determine the angular
velocity and acceleration of the line.
[54.1 rad/s clockwise, 2131 rad/s² anticlockwise]

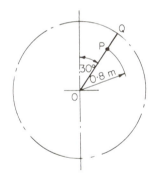

Figure 3.25

9. When a satellite orbits the earth, the only force on the satellite is in the
radial direction, that is, directed towards the earth. The transverse
component of the satellite's acceleration is therefore zero. Use this to
prove Kepler's second law of planetary motion, which states that the
radius vector from the earth to the satellite sweeps through area at a
constant rate.

10. Communications satellites are commonly put into a geostationary orbit,
that is they rotate with the same period as the earth, and thus appear
to be stationary relative to an observer on the earth. What altitude and
what orbital velocity should such a satellite have? The mean radius of
the earth is 6.371×10^8 m, and the acceleration due to gravity at an
altitude h above the earth's surface is $g_0(r^2/[r + h]^2)$ where g_0 is the
value at the earth's surface.
[35 800 km, 11 060 km/h]

11. The compass in a light aircraft shows that it is flying due north. The air speed indicator gives a reading of 150 km/h. If there is a 50 km/h wind blowing from west to east, what is the ground speed of the aircraft?
   [141.4 km/h]

12. A man can row a boat through still water at 4 km/h. How long will it take him to cross an 18 m wide river flow at 2 km/h (a) by rowing at right angles to the current; (b) by rowing in a direction such that he lands directly opposite his starting point?
   [16.2 s, 18.71 s]

13. In figure 3.26 the end of the cable D is pulled off the bottom of the drum with a velocity of 50 mm/s and an acceleration of 20 mm/s², causing the cylinder to roll without slipping. Determine (a) the angular velocity of the drum; (b) the velocity of centre A; (c) the rate of winding or unwinding of the cable on the drum; and (d) the acceleration of points A, B and C.

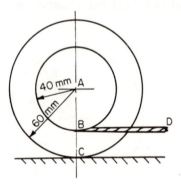

Figure 3.26

   [2.5 rad/s clockwise, 150 mm/s →, 100 mm/s, 60 mm/s² →, 251 mm/s² ∠ 85.4°, 375 mm/s² ↑]

14. Referring to the suspension of figure 2.24 locate the instantaneous centre of the wheel assembly, assuming that the vehicle body moves vertically only, and show that this vertical body movement must be accompanied by horizontal movement (scrub) of the tyre where it contacts the road. Estimate the ratio of the tyre scrub to vertical body movement at the position shown (scrub rate in bounce). Is this ratio constant for all suspension positions? [*Hint*: note that vertical movement of the body while the tyre remains in contact with the road is equivalent to holding the body stationary and moving the road vertically.]
   [Approx. 0.5]

4 Kinetics

4.1 KINETICS OF A PARTICLE

Newton's second law applied to a particle of constant mass m gives

$$R = ma \qquad (4.1)$$

where R is the resultant of the system of forces acting on the particle, and a is the acceleration. This vector relationship is commonly called the *equation of motion* and may be expressed graphically or in scalar components as described in section 2.2.3. In all except the simplest of applications, a free-body diagram will be found indispensable.

Example 4.1

Figure 4.1a shows a particle P of mass 3 kg subjected to four forces. Determine the acceleration of the particle.

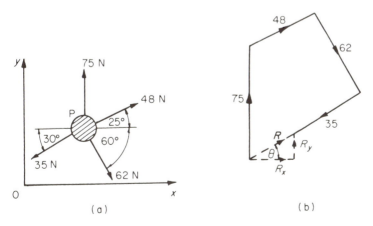

Figure 4.1

Graphical Solution

By drawing the vector diagram, figure 4.1b, the forces are summed to give a resultant

$$R = 50.34 \text{ N}$$

hence the acceleration

$$a = R/m$$
$$= 50.34/3 = 16.78 \text{ m/s}^2$$

at an angle

$$\theta = 28.6°$$

Analytical Solution

Writing the equations of motions in the x- and y-directions

$$48 \cos 25° + 62 \cos 60° - 35 \cos 30° = 3a_x$$
$$75 + 48 \sin 25° - 62 \sin 60° - 35 \sin 30° = 3a_y$$

therefore

$$a_x = 14.73 \text{ m/s}^2$$
$$a_y = 8.03 \text{ m/s}^2$$

hence

$$a = \sqrt{(a_x^2 + a_y^2)} = 16.78 \text{ m/s}^2$$

and

$$\theta = \tan^{-1} a_y/a_x = 28.6°$$

Example 4.2

Figure 4.2a shows a simple device for measuring the effectiveness of a vehicle braking system. The device is carried in the vehicle so that the pendulum can swing in a fore and aft plane. When the vehicle brakes the pendulum swings forwards. Under steady braking conditions, and when oscillations of the pendulum have ceased, the angle θ will be constant and related to the vehicle deceleration. Determine this relationship.

The free-body diagram for the pendulum is shown in figure 4.2b. Since the deceleration (or more appropriately the acceleration) is horizontal, a Cartesian co-ordinate system has been chosen. Writing the equations of

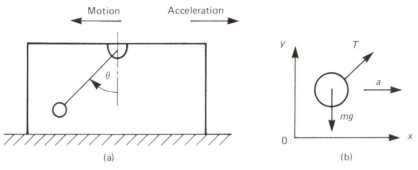

Figure 4.2

motion for the pendulum of mass m

$$T \sin \theta = ma$$
$$T \cos \theta - mg = 0$$

Hence

$$a = g \tan \theta$$

Numerically, of course, the deceleration and the acceleration are equal.

Example 4.3

In figure 4.3a the particle P is a distance r from O when the tube OA starts to rotate from rest anticlockwise in a horizontal plane with a constant angular acceleration of 4.8 rad/s². If the coefficient of friction between the particle and the tube is 0.3, at what tube angular velocity will the particle start to slide? In which direction will sliding commence?

Figure 4.3b shows the free-body diagram for the particle together with the r–θ acceleration components. Until sliding starts, the particle moves in

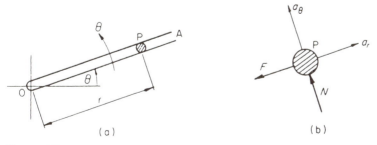

Figure 4.3

a circle so that r is a constant and hence

$$a_r = -r\dot{\theta}^2$$

$$a_\theta = r\ddot{\theta}$$

The minus sign for a_r indicates that the radial acceleration is towards O and hence the radial force due to friction must also be directed towards O. Since friction opposes the tendency to move, the particle will ultimately slide outwards.

The equations of motion give

$$F = -mr\dot{\theta}^2$$

$$N = mr\ddot{\theta} = mr \times 4.8$$

but, for no sliding

$$F/N \leqslant \mu = 0.3$$

therefore

$$\dot{\theta} \leqslant \sqrt{(0.3 \times 4.8)} = 1.2 \text{ rad/s}$$

hence sliding will start when the angular velocity exceeds 1.2 rad/s. Note that this is independent of the position of the particle in the tube.

4.2 KINETICS OF A RIGID BODY

Consider a system of particles forming a rigid body, figure 4.4a, subjected to a number of external forces of which only three, F_1, F_2 and F_3 are shown. A typical particle of mass m_i is located at the point (x_i, y_i), and the point G having co-ordinates x_G and y_G is the mass centre of the system. Figure 4.4b is a free-body diagram for the particle. Some of the external forces, F_{i1}, F_{i2} and F_{i3} may act directly on m_i, and there will also be internal forces or reactions from adjacent particles. The resultant of all these forces will give

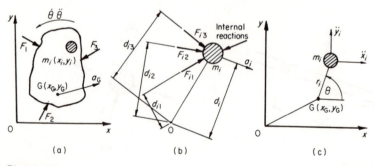

(a)   (b)   (c)

Figure 4.4

an acceleration a_i, so that by applying equation 4.1

$$F_{i1} + F_{i2} + F_{i3} + (\text{reactions from adjacent particles}) = m_i a_i.$$

Summing for all particles, the internal reactions will cancel, leaving

$$\Sigma F = R = \Sigma m_i a_i$$

or in scalar form

$$\Sigma F_x = R_x = \Sigma m_i \ddot{x}_i$$

$$\Sigma F_y = R_y = \Sigma m_i \ddot{y}_i$$

Now from the definition of mass centre (section 2.1.3)

$$mx_G = \Sigma m_i x_i$$

$$my_G = \Sigma m_i y_i$$

Differentiating twice

$$m\ddot{x}_G = \Sigma m_i \ddot{x}_i$$

$$m\ddot{y}_G = \Sigma m_i \ddot{y}_i$$

so that

$$\Sigma F_x = m\ddot{x}_G \tag{4.2}$$

$$\Sigma F_y = m\ddot{y}_G \tag{4.3}$$

These two linear equations of motion may be combined and expressed in vector form

$$\Sigma F = ma_G \tag{4.4}$$

Note that this result is *not dependent on any angular motion* that the body may have. To analyse this angular motion consider the moments of the forces on the particle about the origin O.

$$M_O = F_{i1} d_{i1} + F_{i2} d_{i2} + F_{i3} d_{i3} + (\text{moments of internal reactions})$$

$$= m_i a_i d_i$$

Resolving a_i and d_i into cartesian components (figure 4.4c) and summing for all particles

$$\Sigma M_O = \Sigma m_i (\ddot{y}_i x_i - \ddot{x}_i y_i) \tag{4.5}$$

Using the principle of relative motion, section 3.3

$$x_i = x_{i/G} + x_G$$

$$\ddot{x}_i = \ddot{x}_{i/G} + \ddot{x}_G$$

$$y_i = y_{i/G} + y_G$$

$$\ddot{y}_i = \ddot{y}_{i/G} + \ddot{y}_G$$

Substituting these terms into equations 4.5 and expanding

$$\Sigma M_O = \Sigma m_i x_{i/G} \ddot{y}_{i/G} + x_G \Sigma m_i \ddot{y}_{i/G} + \ddot{y}_G \Sigma m_i x_{i/G} + x_G \ddot{y}_G \Sigma m_i$$
$$- \Sigma m_i y_{i/G} \ddot{x}_{i/G} - y_G \Sigma m_i \ddot{x}_{i/G} - \ddot{x}_G \Sigma m_i y_{i/G} - y_G \ddot{x}_G \Sigma m_i \qquad (4.6)$$

Now from section 2.1.3 the terms $\Sigma m_i x_{i/G}$ and $\Sigma m_i y_{i/G}$ are both zero and by differentiating it is clear that the terms $\Sigma m_i \ddot{x}_{i/G}$ and $\Sigma m_i \ddot{y}_{i/G}$ are also zero. In addition

$$x_{i/G} = r_i \cos \theta$$

$$y_{i/G} = r_i \sin \theta$$

therefore

$$\ddot{x}_{i/G} = -r_i \ddot{\theta} \sin \theta - r_i \dot{\theta}^2 \cos \theta$$

$$\ddot{y}_{i/G} = r_i \ddot{\theta} \cos \theta - r_i \dot{\theta}^2 \sin \theta$$

By substitution into equation 4.6 the moment sum about O reduces to

$$\Sigma M_O = m\ddot{y}_G x_G - m\ddot{x}_G y_G + \ddot{\theta} \Sigma m_i r_i^2$$

The quantity $\Sigma m_i r_i^2$ is defined as the moment of inertia of the body about G and is given the symbol I_G, thus

$$\Sigma M_O = m\ddot{y}_G x_G - m\ddot{x}_G y_G + I_G \ddot{\theta} \qquad (4.7)$$

In any particular problem equation 4.7 may be used directly (together with equations 4.2 and 4.3, or equation 4.4 as appropriate) choosing a convenient point O about which to consider the moment sum. However, in practice, the application is often made simpler by restricting the choice according to the classification of motion described in section 3.2, namely, translation, fixed-axis rotation and general plane motion. These three cases will now be considered.

4.2.1   Translation of a Rigid Body

In this case, whether the motion is rectilinear or curvilinear, there is no rotation, so that $\ddot{\theta}$ is zero. If in addition the point O is taken as the mass centre, that is, G, then x_G and y_G are both zero, so that from equation 4.7

$$\Sigma M_G = 0 \qquad (4.8)$$

Example 4.4

The mass centre of a motorcycle and rider is mid-way between the wheels which are 1.8 m apart, and 1.2 m above the road surface. Assuming adequate engine torque, what is the maximum acceleration that can be obtained along a straight level road if the coefficient of friction μ between the rear driving wheel and the road is 0.8? Would an increase in μ enable a higher acceleration to be obtained?

Figure 4.5 shows the free-body diagram. The moment of inertia of the wheels has been neglected so that the reaction at the front wheel is vertical.

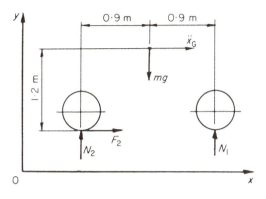

Figure 4.5

Taking moments about the mass centre, equation 4.8 gives

$$0.9\,(N_2 - N_1) - 1.2F_2 = 0$$

The maximum acceleration will occur when either the front wheel lifts or the rear wheel skids. Assume the former so that N_1 is zero giving

$$F_2/N_2 = 0.9/1.2 = 0.75$$

but

$$F_2/N_2 \leqslant \mu = 0.8$$

hence the assumption is correct and limiting friction has not been reached. Summing the forces horizontally and vertically, equations 4.2 and 4.3 give

$$F_2 = m\ddot{x}_G$$

$$N_2 - mg = 0$$

hence

$$\ddot{x}_G = F_2/m = 0.75N_2/m = 0.75g = 7.36 \text{ m/s}^2$$

Since limiting friction has not been reached an increase in μ will have no effect on the maximum obtainable acceleration.

Note that having established that the front-wheel reaction is zero, the maximum acceleration can be obtained directly from a moment sum about point A, using equation 4.7 as follows

$$\Sigma M_A \circlearrowleft \quad 0.9mg - 1.2m\ddot{x}_G = 0$$

hence

$$\ddot{x}_G = (0.9/1.2)g = 7.36 \text{ m/s}^2$$

Example 4.5

Determine the acceleration of the 20 kg mass in each case in figure 4.6a and b. The moment of inertia of the pulley is negligible, and there are no significant resistances, so that the tension in the cable on one side of the pulley is equal to that on the other.

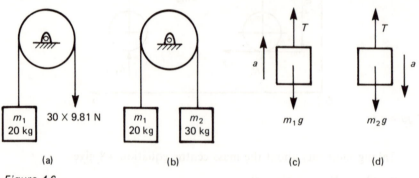

Figure 4.6

The free-body diagram for the 20 kg mass, figure 4.6c, will be the same for both cases, although, as will be seen, the value of the tension T and acceleration a will differ. Figure 4.6d shows the free-body diagram for the 30 kg mass which is only relevant for the second case. It is intuitively obvious that the accelerations of the two masses in the second case will be equal and opposite.

(a)

$$T = 30 \times 9.81$$

$$T - m_1 g = m_1 a$$

Hence

$$a = 9.81(30 - 20)/20 = 4.91 \text{ m/s}^2$$

(b) As before

$$T - m_1 g = m_1 a$$

In addition

$$m_2 g - T = m_2 a$$

Adding to eliminate T

$$a = 9.81(30 - 20)/(30 + 20) = 1.96 \text{ m/s}^2$$

Comparison of the denominators in the solution for a shows clearly why there is a difference.

4.2.2 Fixed-axis Rotation of a Rigid Body

For this type of motion it is usually convenient to put the point O at the axis of rotation, figure 4.7, so that equation 4.7 becomes

$$\Sigma M_O = m\ddot{y}_G x_G - m\ddot{x}_G y_G + I_G \ddot{\theta}$$
$$= mr \cos \theta(-r\dot{\theta}^2 \sin \theta + r\ddot{\theta} \cos \theta)$$
$$- mr \sin \theta(-r\dot{\theta}^2 \cos \theta - r\ddot{\theta} \sin \theta) + I_G \ddot{\theta} = (I_G + mr^2)\ddot{\theta}$$

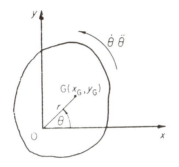

Figure 4.7

The term $(I_G + mr^2)$ is the *moment of inertia of the body about O*, I_O, so that

$$\Sigma M_O = I_O \ddot{\theta} \tag{4.9}$$

The expression

$$I_O = I_G + mr^2$$

shows the use of the *parallel-axis theorem*. Writing I_G in terms of the centroidal radius of gyration, k_G

$$I_O = m(k_G^2 + r^2) = mk_O^2$$

thus defining the *radius of gyration about point O*.

Example 4.6

A rigid body having a centroidal radius of gyration k_G is pivoted about a point O at a distance l from the mass centre and allowed to oscillate through a small angle. Show that the motion is simple harmonic and hence find an expression for the time of one complete oscillation. For what value of l will this time be a minimum?

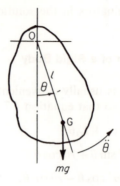

Figure 4.8

Figure 4.8 shows the body having been displaced through a small angle θ, with the angular acceleration $\ddot{\theta}$ *measured positive in the same direction.* From equation 4.9

$$-mgl \sin \theta = I_0 \ddot{\theta}$$

hence for small angles

$$\ddot{\theta} = -(mgl/I_0)\theta$$

From example 3.4 this is seen to be the condition for simple harmonic motion and the solution for the displacement is

$$\theta = \frac{\omega_0}{\sqrt{k}} \sin \sqrt{k}\, t$$

where $k = (mgl/I_0)$ and ω_0 is the initial angular velocity. For one complete oscillation, $\sqrt{k}\,t$ is 2π and hence

$$t = 2\pi/\sqrt{k} = 2\pi\sqrt{(I_0/mgl)}$$

Using the parallel-axis theorem this becomes

$$t = 2\pi \sqrt{\left(\frac{k_G^2 + l^2}{gl}\right)}$$

For a stationary value

$$\frac{dt}{dl} = 0 = \frac{1}{2}\left(\frac{k_G^2 + l^2}{gl}\right)^{-\frac{1}{2}}\left[\frac{gl \times 2l - (k_G^2 + l^2)g}{(gl)^2}\right]$$

hence

$$2l^2 = k_G^2 + l^2$$

therefore

$$l = k_G$$

If the pivot point O is at G, then there can be no moment about O and the time for one oscillation will be infinite. Since there is only one solution, it follows that it must define a minimum time.

Table 4.1 lists the expressions for the moments of inertia of some common bodies.

Table 4.1   Moments of Inertia

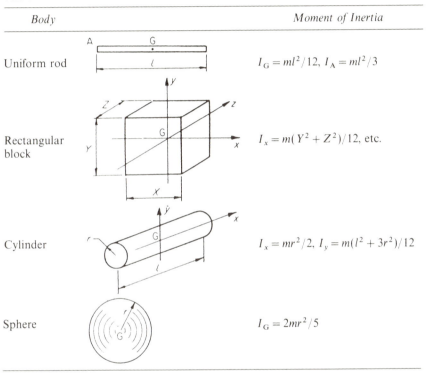

Body	Moment of Inertia
Uniform rod	$I_G = ml^2/12, \ I_A = ml^2/3$
Rectangular block	$I_x = m(Y^2 + Z^2)/12$, etc.
Cylinder	$I_x = mr^2/2, \ I_y = m(l^2 + 3r^2)/12$
Sphere	$I_G = 2mr^2/5$

Example 4.7

If, in example 4.5, the pulley has a mass of 10 kg and a radius of 100 mm, what will be the acceleration of the 20 kg mass for the second case, figure 4.6b? Treat the pulley as a uniform disc and use table 4.1.

Since the pulley has inertia the tensions will not be equal, and three free-body diagrams must be considered, one for each body having mass, figure 4.9. For completeness, the pulley bearing reaction R and the pulley weight $m_p g$ have been shown, but they play no part in the solution.

Application of the appropriate equations of motion gives

$$T_1 - m_1 g = m_1 a$$
$$m_2 g - T_2 = m_2 g$$
$$(T_2 - T_1)r = I\alpha$$

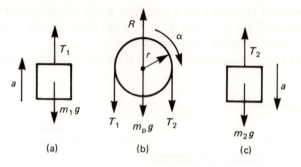

Figure 4.9

From table 4.1

$$I = m_p r^2/2 = 10 \times 0.1^2/2 = 0.05 \text{ kgm}^2$$

and from a consideration of the kinematics of the system

$$a = r\alpha$$

Substituting for α and eliminating T_1 and T_2 gives

$$a = g(m_2 - m_1)/(m_1 + m_2 + I/r^2)$$
$$= 9.81(30 - 20)/(30 + 20 + 0.05/0.1^2)$$
$$= 1.78 \text{ m/s}^2$$

4.2.3   General Plane Motion of a Rigid Body

In this case it is generally more convenient to consider a moment sum about the mass centre. Equation 4.7 thus becomes

$$\Sigma M_G = I_G \ddot{\theta} \qquad\qquad (4.10)$$

Example 4.8

A uniform cylinder of radius 40 mm is released from rest on an incline of 30°. Find the acceleration of the centre if the coefficient of friction is (a) 0.3, (b) 0.15, (c) zero.

Referring to figure 4.10 and applying equations 4.2, 4.3 and 4.10

$$mg \sin 30° - F = ma$$
$$N - mg \cos 30° = 0$$
$$Fr = (mr^2/2)\alpha$$

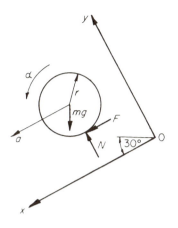

Figure 4.10

Also, using the result of example 3.11 assuming pure rolling

$$a = r\alpha$$

therefore

$$F = mg \sin 30° - mr(2Fr/mr^2)$$

giving

$$F = \tfrac{1}{3}mg \sin 30°$$

But

$$N = mg \cos 30°$$

therefore

$$F/N = \tfrac{1}{3}\tan 30° = 0.192$$

For (a) $F/N < \mu$ hence pure rolling does occur and limiting friction has not been reached. Therefore

$$mg \sin 30° = \tfrac{1}{2}mr\frac{a}{r} = ma$$

giving

$$a = \tfrac{2}{3}g \sin 30° = 3.27 \text{ m/s}^2$$

For (b) it is clear that pure rolling is not possible so that

$$a \neq r\alpha$$

but since limiting friction applies

$$F/N = \mu$$

and the acceleration becomes

$$a = g(\sin 30° - \mu \cos 30°) = 3.63 \text{ m/s}^2$$

For (c), if μ is zero so also is F, and the cylinder slides down without rotation giving

$$a = g \sin 30° = 4.90 \text{ m/s}^2$$

Example 4.9

Using the data from example 3.13 determine the tension in each cable if the mass of the load is 102 kg and its centroidal moment of inertia is 200 kgm^2.

The equations of motion for the load are

$$T_A + T_B - mg = ma_G$$
$$1.5 \, T_B - 1.5 \, T_A = I_G \ddot{\theta}$$

But from example 3.13 a_G is 0.125 m/s^2 and $\ddot{\theta}$ is 0.017 rad/s^2, giving

$$T_A = 505 \text{ N}$$
$$T_B = 508 \text{ N}$$

4.3   D'ALEMBERT'S PRINCIPLE

In the earlier section on kinetics of a particle, problems were solved by summing the external forces to give a resultant, and equating this resultant to the product of mass and acceleration. This approach is straightforward and comes directly from Newton's second law. However, if the basic equation for the motion of a particle, equation 4.1, is written

$$\Sigma F - ma = 0$$

and the quantity *ma* is treated as a force in the same way as those contained in the summation ΣF, then the problem appears to be one of statics. The way this is done in practice is to apply a fictitious force of magnitude *ma* (often called an *inertia force*) to the accelerating particle, but *opposite in direction to the acceleration*, and then to solve the problem using the conditions for equilibrium discussed in section 2.3. This *apparent* transformation of a problem of kinetics to one in statics is known as *D'Alembert's principle*. It is easily extended to the case of a rigid body by applying the inertia force through the mass centre and by adding an inertia couple of magnitude $I_G \alpha$ but opposite in direction to the angular acceleration. Equation 4.7 for example would then be written

$$\Sigma M_O - m\ddot{y}_G x_G + m\ddot{x}_G y_G - I_G \ddot{\theta} = 0 \qquad (4.11)$$

Example 4.10

The mass of the rod in example 3.12 is 4 kg and the coefficient of friction between the ends A and B and their respective guides along the *y*- and *x*-axes is 0.5. Determine the horizontal force *P* that must be applied at B to produce the motion, if the rod moves in a vertical plane.

Figure 4.11 shows the free-body diagram for the rod together with the components of the inertia force and the inertia couple. From equation 4.11 about point B

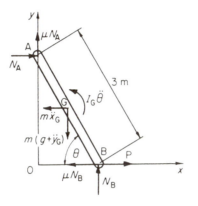

Figure 4.11

$$3N_A \sin \theta + 3\mu N_A \cos \theta - 1.5m(g + \ddot{y}_G) \cos \theta - 1.5m\ddot{x}_G \sin \theta - I_G \ddot{\theta} = 0$$

but from example 3.12

$$\ddot{x}_G = 0.25 \text{ m/s}^2, \ \ddot{y}_G = -1.03 \text{ m/s}^2, \ \ddot{\theta} = -0.535 \text{ rad/s}^2$$

hence for $\theta = 60°$

$$N_A = 7.80 \text{ N}$$

Vertically

$$N_B - m(g + \ddot{y}_G) + \mu N_A = 0$$

therefore

$$N_B = 31.2 \text{ N}$$

Horizontally

$$P - \mu N_B - m\ddot{x}_G + N_A = 0$$

hence

$$P = 8.80 \text{ N}$$

Opinion on the merits of D'Alembert's principle differ widely. It is the authors' opinion that while its use may be helpful in certain types of problem, the more direct approach should be used until the student has obtained a thorough understanding of the fundamental relationship between force, mass and acceleration.

SUMMARY

For a particle

$$\Sigma F = R = ma \qquad \text{or} \qquad \begin{aligned} \Sigma F_x &= R_x = m\ddot{x} \\ \Sigma F_y &= R_y = m\ddot{y} \end{aligned}$$

For a rigid body

$$\Sigma F = R = ma_G \qquad \text{or} \qquad \begin{aligned} \Sigma F_x &= R_x = m\ddot{x}_G \\ \Sigma F_y &= R_y = m\ddot{y}_G \end{aligned}$$

$$\Sigma M_O = m\ddot{y}_G x_G - m\ddot{x}_G y_G + I_G\ddot{\theta}$$

where O is any point and G is the mass centre.

In particular

$$\Sigma M_G = I_G\ddot{\theta}$$

Special cases

Translation

$$\Sigma M_O = 0$$

Fixed axis rotation about some point O

$$\Sigma M_O = I_O\ddot{\theta}$$

D'Alembert's principle

$$\Sigma M_O - m\ddot{y}_G x_G + m\ddot{x}_G y_G - I_G\ddot{\theta} = 0$$

4.4   PROBLEMS

1. A 1500 kg elevator ascends 25 m in 6 s. The motion consists of a steady acceleration from rest followed by a period of a constant velocity of 5 m/s and finally a constant deceleration to rest. If the tension in the supporting cable during acceleration is 25.2 kN, what is the tension during deceleration?
   [8.85 kN]

2. Referring to the lawn sprinkler of example 3.9, if the internal diameter of the tubular arms is 25 mm, what torque is required to drive the system? The density of water is 1000 kg/m³.
   [0.556 Nm]

3. The piston of an IC engine has a diameter of 100 mm and a mass of 1.6 kg. At a particular point in the cycle, the cylinder pressure is 1.1 N/mm² and the piston is moving upwards with an acceleration of 2860 m/s². If the normal reaction between the piston and cylinder wall is 2000 N and the coefficient of friction is 0.17, determine the force exerted on the piston by the connecting-rod.
   [13.7 kN, 8.38° to line of stroke]

4. Figure 4.12 shows a very simple but surprisingly accurate device for giving a single point calibration for an accelerometer. The accelerometer A is securely attached to the plate which vibrates vertically with simple harmonic motion. The steel ball B rests on the plate but is not attached to it. It will be found that as the frequency of vibration is increased, for a given displacement amplitude, there will come a point where the ball will be heard to 'bounce'. When will this occur? If the displacement amplitude is 2 mm, at what frequency will bounce begin?
   [11.15 Hz]

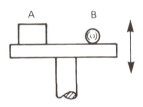

Figure 4.12

5. A car rounds a circular bend at a constant speed, the road being banked inwards. The mass centre of the car is 0.8 m above the ground, and the transverse distance between the wheels is 1.2 m. If the speed is sufficiently high, the car will either roll over or skid. Which will occur first if the coefficient of friction between the road and the wheels is 0.85? If the angle of bank is 3° and the radius of turn is 40 m, at what speed will roll-over or skid commence?
   [Roll-over at 65.2 km/h]

6. The maximum acceleration that can be attained by a motorcycle on level ground will be limited by the engine torque, by lifting of the front wheel, or by skidding of the rear driving wheel. In a particular case the mass of the motorcycle and rider is 200 kg with the mass centre 1.2 m above ground level and midway between the wheels which are 1.8 m apart. The maximum available rear-wheel driving torque is 600 Nm, the rear-wheel diameter is 0.85 m and the coefficient of friction between the tyre and the road is 0.7. What is the maximum possible acceleration?
   [6.44 m/s²]

7. In example 2.6 determine the acceleration of the platform if the torque M is 500 Nm, and the link AB has an angular velocity of 0.75 rad/s anticlockwise. Ignore the masses of the links.

   [1.84 m/s², ↘ 44.2°]

8. A constant torque of 150 Nm applied to a turbine rotor is sufficient to overcome the constant bearing friction and to give it a speed of 75 rev/min from rest after 9 revolutions. When the torque is removed, the rotor turns for a further 23 revolutions before stopping. Determine the moment of inertia of the rotor and the bearing friction.

   [198 kg m², 42.2 Nm]

9. An electric hoist consists of a motor coupled directly to a pinion meshing with a gear wheel on a cable drum from which is supported a load, as shown in figure 4.13. The moment of inertia of the rotating parts of the motor and pinion are 25 kgm² and that of the gear wheel and drum is 85 kgm². The load has a mass of 200 kg. The pinion and wheel have effective radii of 120 and 620 mm respectively. What motor torque is required to raise the load with a constant acceleration of 0.6 m/s².

   [282.2 Nm]

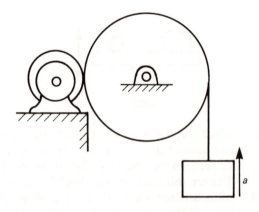

Figure 4.13

10. A uniform vertical bar is pivoted about a horizontal axis through its lower end. It is given a small angular displacement so that it begins to swing downwards starting from rest. If the mass of the bar is 150 kg determine the reaction on the bar at the pivot when it has turned through 20° from the vertical.

    [1256 N, ∠ 75.8°]

11. A disc, a sphere and a thin-walled tube are placed on an incline and roll down without slip. Which will acquire the greatest acceleration? Does the answer depend on the objects having the same diameter?

    $[a_{\text{sphere}} > a_{\text{disc}} > a_{\text{tube}}]$

12. A uniform bar is supported horizontally by two vertical cables, one at each end. If one cable suddenly breaks, find the initial tension of the other in terms of the tension before breaking.
    $[\frac{1}{2}]$

13. Solve the previous problem if the upper ends of the two cables are connected to a common point directly above the centre of the bar and subtend an included angle of 2ϕ. [*Hint*: consider the acceleration of one end of the bar relative to the mass centre.]
    $[2\cos^2\phi/(3\cos^2\phi+1)]$

5 Work and Energy

5.1 THE NATURE OF WORK

When applying Newton's second law of motion, the forces and accelerations are those at an instant, and thus give instantaneous solutions; however, in many cases the relationship between the forces and the resulting change of motion over a finite interval needs to be considered. The concept of *work*, together with the closely associated concept of *energy*, give a method for doing this where the interval is one of displacement.

In everyday life work is associated with the expenditure of physical or mental effort. In dynamics, the term work has a more precise definition as described in sections 5.1.1 and 5.1.2, and these should be studied carefully, since the definition sometimes appears to be at variance with everyday experience.

5.1.1 Work Done by a Force

Consider a force F applied to a particle or rigid body, figure 5.1a, and suppose that the point of application of F moves a small distance δs during which the force can be assumed to remain constant. The work done by F during this displacement is defined as

$$\delta W = F \cos \theta \, \delta s$$

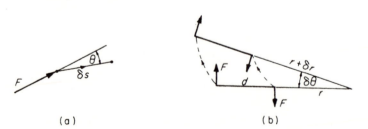

(a)   (b)

Figure 5.1   *Work done by a force and a couple*

Note that this can be considered as either

(a) the product of δs and the component of F in the direction of δs, or
(b) the produce of F and the component of δs in the direction of F.

By summing all such small displacements over a finite distance, the work done is

$$W = \int_0^s F \cos \theta \, \mathrm{d}s \qquad (5.1)$$

and is measured in newton metres or *joules*.

Work in a scalar quantity but if $\theta > 90°$ work is negative, indicating work done on the force. For example, limiting friction will always result in negative work. Also, because work is a scalar, the total work done by a system of forces is merely the algebraic sum of the work done by each individual force.

A situation often arises where a force acts at right angles to the motion of its point of application, and such forces will, of course, do no work. The normal reactions between the road and the wheels of a vehicle are good examples of this.

Another example arises when a man walks carrying a heavy load on his shoulders. By the definition of equation 5.1 the man does no work on the load, but it would be difficult to convince him of this! The apparent contradiction can be resolved by considering work done internal to the man, but this is beyond the scope of this book.

5.1.2 Work Done by a Couple

Consider the couple shown in figure 5.1b, which translates and rotates by amounts defined by δr and $\delta \theta$. The work done during the small displacement (ignoring second-order terms) is

$$\delta W = F(d + r)\delta \theta - Fr\,\delta \theta$$

$$= Fd\,\delta \theta$$

and therefore the total work done during a finite displacement is

$$W = \int Fd\,\mathrm{d}\theta = \int M\,\mathrm{d}\theta \qquad (5.2)$$

where M is the moment of the couple.

5.2 THE NATURE OF ENERGY

When work is done on a system, some of it is stored inside the system. In this stored form the work is known as *energy*, and some of this energy may be recovered so that the system can do useful work. Consider a couple applied to a key used to wind a clock. Some of the work will be expended in overcoming friction and this part is often described as being *lost* in the sense that it does not represent useful work, but a substantial proportion will be stored in the clock spring, and will be released over a rather longer period of time to drive the mechanism. Of course, friction must still be overcome during the release of energy, and represents a further loss. In practice, whenever work is stored as energy, or energy is recovered as work, a certain amount will be lost from the system and will ultimately manifest itself as heat, which then raises the temperature of the environment.

Energy has many forms, for example chemical energy exists in the form of fuel for road vehicles, nuclear energy locked up in the atomic structure is used for power generation, and electrical energy is widely used in the form of batteries. In dynamics we recognise three forms of energy, namely, potential, strain and kinetic. These are considered separately in sections 5.2.1 to 5.2.3, although it should be noted that some texts treat potential and strain energy together.

5.2.1 Potential Energy (PE)

When work is expended raising a body of mass m slowly from a height h_1 to a height h_2, the vertical force on the body must be equal to its weight, so that from equation 5.1

work done = potential energy gained due to gravity

$$= mg(h_2 - h_1) \tag{5.3}$$

Note that this is independent of any horizontal displacements that may occur, and since problems almost always involve consideration of the change of energy, the heights h may be measured from any convenient datum.

5.2.2 Strain Energy (SE)

If a force F is applied to an elastically deformable body such that the point of application of F moves a distance dx in the direction of F, then, by definition, the work done by F is

$$dW = F\,dx$$

and over a finite displacement, the total work done is

$$W = \int_{x_1}^{x_2} F\,dx$$

and is called the *gain in strain energy*, figure 5.2a.

In dynamics, most of the elastic bodies of interest are in the form of springs, and these usually have a linear stiffness k (force per unit deflection)

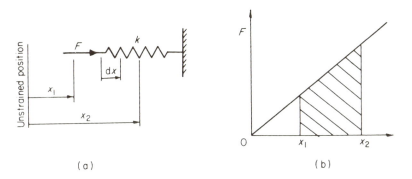

Figure 5.2 Strain energy

so that $F = kx$, figure 5.2. The expression for the gain in strain energy for a linear spring then becomes

$$W = \int_{x_1}^{x_2} F \, dx = \int_{x_1}^{x_2} kx \, dx = \tfrac{1}{2}k(x_2^2 - x_1^2)$$ (5.4)

which is represented by the shaded area in figure 5.2b.

5.2.3 Kinetic Energy of a Particle (KE)

Consider the work required to accelerate a particle from an initial velocity v_1 to a velocity v_2. A force F applied to a particle of mass m gives an acceleration a, so that

$$F = ma = mv \, dv/ds$$

During a displacement s the work done is

$$\int F \, ds = m \int_{v_1}^{v_2} v \, dv = \tfrac{1}{2}m(v_2^2 - v_1^2)$$ (5.5)

and is called the *gain in kinetic energy*.

Examination of the expressions for the three forms of mechanical energy confirms that the units are the same as for work, that is Nm or joules.

5.3 THE WORK–ENERGY EQUATION

The way in which work can be stored as energy and recovered as work, with losses at each stage, has already been described. This leads to the fundamental *energy equation* or *first law of thermodynamics*

Work done on a system = gain of energy + heat lost

which, in dynamics, reduces to

Work done on a system = gain of (PE + SE + KE) + losses   (5.6)

with the further refinement that losses due to Coulomb friction are often most conveniently considered as negative work and included in the left-hand side of equation 5.6.

Example 5.1

How much useful work has been done in getting a 9500 kg aircraft to an altitude of 2000 m and a speed of 500 km/h?

Useful work done = Gain of energy (PE and KE in this case)

From equation 5.3

Gain of PE = $9500 \times 9.81 \times 2000 = 186.4$ MJ

From equation 5.5

Gain of KE = $\frac{1}{2} \times 9500 \times (500/3.6)^2 = 91.6$ MJ

Hence

Work done = $186.4 + 91.6 = 278.0$ MJ

Example 5.2

A 12 kg package is given an initial velocity of 2 m/s down a 30° ramp, as shown in figure 5.3, and strikes the spring which has a stiffness of 1500 N/m and an initial compression of 130 mm. If the coefficient of friction between the package and the ramp is 0.2, determine the maximum compression of the spring.

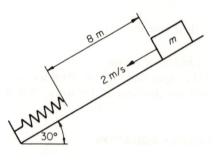

Figure 5.3

Let the maximum compression be x. This will occur at the instant when the package is brought to rest.

work done against friction = $-\mu mg \cos 30°(8 + x - 0.13)$

gain of PE $= -mg(8 + x - 0.13)\sin 30°$

gain of SE $= \frac{1}{2}k(x^2 - 0.13^2)$

gain of KE $= -\frac{1}{2}m \times 2^2$

Substituting into the energy equation 5.6 and simplifying

$x^2 - 0.0513x - 0.452 = 0$

and taking the positive root

$x = 0.698$ m

Note that only the initial and final states have been considered, in spite of the discontinuity in the motion.

5.4   KINETIC ENERGY OF A RIGID BODY

Figure 5.4 shows a rigid body with a typical particle m_i at a distance r_i from the mass centre G. The velocity of the particle is v_i so that its kinetic energy is $\frac{1}{2}m_i v_i^2$, and thus summing for all particles, the total KE is $\Sigma \frac{1}{2}m_i v_i^2$. Using the principle of relative motion

$$v_i = v_{i/G} + v_G = r_i \omega + v_G$$

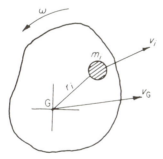

Figure 5.4   *Kinetic energy of a rigid body*

hence

$$KE = \omega^2 \Sigma \frac{1}{2}m_i r_i^2 + v_G^2 \Sigma \frac{1}{2}m_i + \omega v_G \Sigma m_i r_i$$

$$= \frac{1}{2}I_G \omega^2 + \frac{1}{2}mv_G^2 \tag{5.7}$$

If the body is rotating about a fixed axis O at a distance r from G, then

$$v_G = r\omega$$

and hence

$$KE = \frac{1}{2}I_G \omega^2 + \frac{1}{2}mr^2 \omega^2 = \frac{1}{2}I_O \omega^2 \tag{5.8}$$

This will also hold true where O is the instantaneous centre of zero velocity (see section 3.3.1).

Example 5.3

The uniform rectangular trap-door shown in figure 5.5 has a mass of 9 kg, and is released from rest in the horizontal position. If there is a frictional torque of 23 Nm at the pivot O, determine the angular velocity of the door just before it strikes the rubber stop at A. If after rebound, the door rises to the position where θ is 20°, find how much energy has been lost from the system due to the inelasticity of the rubber stop.

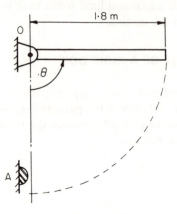

Figure 5.5

Applying equation 5.6 from the position $\theta = 90°$ to just before impact

$$-23\frac{\pi}{2} = -mg \times 0.9 = \frac{1}{2}\left(m \times \frac{1.8^2}{3} \right)\omega^2$$

giving

$$\omega = 2.99 \text{ rad/s}$$

Considering now the interval from $\theta = 90°$ to $\theta = 20°$ after rebound

$$-23\left(\frac{\pi}{2} + \frac{\pi}{9} \right) = -mg \times 0.9 \times \cos 20° + \text{losses}$$

hence

$$\text{losses} = 30.5 \text{ J}$$

Example 5.4

A disc of mass m is placed so that it can roll along a horizontal surface and a couple M, equal to the weight of the disc multiplied by its radius, is applied in its plane. Determine the total kinetic energy of the disc and the work done by the couple when the disc has acquired a velocity v from rest (a) if the coefficient of friction is 0.75; (b) if the coefficient of friction is 0.5.

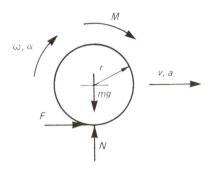

Figure 5.6

Figure 5.6 shows the free-body diagram for the system. Clearly, the weight mg and normal reaction N will not do any work, but the friction force F may or may not depending on whether or not slip occurs. This must be considered first.

Summing the forces horizontally

$$F = ma$$

Taking moments about the mass centre O

$$M - FR = I\alpha = (mr^2/2)\alpha$$

But

$$M = mgr$$

and if no slip is assumed

$$a = r\alpha$$

Solving these equations for F

$$F = 2mg/3$$

hence

$$F/N = 2/3$$

therefore, for no slip

$$\mu > 2/3$$

It is clear that slip does not occur in the first part of the problem ($\mu = 0.75$) but it does in the second ($\mu = 0.5$).

(a) *No slip*

Since there are no losses

$$\text{Work done by } M = M\theta = \text{Total KE} = \tfrac{1}{2}mv^2 + \tfrac{1}{2}I\omega^2$$

For rolling without slip

$$V = r\omega$$

hence

$$M\theta = \tfrac{1}{2}mv^2 + \tfrac{1}{2}(mr^2/2)(v/r)^2 = 3mv^2/4$$

(b) *Slip occurs*

In addition to generating the kinetic energy, the couple must do sufficient work to overcome the losses, that is, the work done against friction. Thus

$$M\theta = \tfrac{1}{2}mv^2 + \tfrac{1}{2}(mr^2/2)\omega^2 + \text{WD against friction}$$

The first two terms give the total kinetic energy, but the linear and angular velocities, v and ω, are no longer simply related. The necessary relationships between angular and linear motion can be determined by using the principles of chapter 4, noting that we can additionally use the limiting friction condition. Thus

$$F = \mu N = \mu mg = ma$$

giving

$$a = \mu g = g/2$$

and

$$M - FR = mgr - \mu mgr = I\alpha = (mr^2/2)\alpha$$

giving

$$\alpha = 2g(1 - \mu)/r = g/r$$

thus

$$\alpha/a = 2/r$$

But since the disc accelerates uniformly from rest, there is a constant relationship between angular and linear motion, that is

$$\alpha/a = 2/r = \omega/v = \theta/x$$

where θ is the angular displacement and x is the linear displacement of the centre of the disc.

In addition

$$v^2 = 0 + 2ax$$

giving

$$x = v^2/2a = v^2/g$$

Then

$$\text{Total KE} = \tfrac{1}{2}mv^2 + \tfrac{1}{2}(mr^2/2)(2v/r)^2 = 3mv^2/2$$

hence

$$\text{WD against friction} = F \times \text{slipping distance}$$
$$= \mu mg(r\theta - x)$$
$$= \tfrac{1}{2}mgx = mv^2/12$$

and

$$\text{WD by } M = \text{KE} + \text{WD against friction}$$
$$= 2mv^2$$

5.5 POWER

When discussing work and energy in the previous section, the winding of a clock spring and subsequent running down was given as an example. In this case the work is put into the system relatively quickly, say in a few seconds, and taken out rather more slowly, perhaps in hours or even days. On the other hand when firing an arrow, the work is put in slowly compared with the very rapid discharge. In each case the total work into the system equals the work out (ignoring losses) but the *rate of doing work*, defined as the *power*, is very different. From this definition and equation 5.1

$$\text{power} = \frac{\mathrm{d}}{\mathrm{d}t}\left[\int_0^s F\cos\theta\,\mathrm{d}s \right] = F\cos\theta\,\frac{\mathrm{d}s}{\mathrm{d}t} = F\cos\theta\,v \tag{5.9}$$

Similarly, for a couple, the power developed is

$$P = M\,\mathrm{d}\theta/\mathrm{d}t = M\omega \tag{5.10}$$

Since work equals the gain of energy, power can sometimes be obtained more conveniently by putting

$$P = \text{rate of gain of energy}$$

Power is measured in units of newton metres/second or *watts*.

Example 5.5

The resistance to motion of a 20 000 Mg ship is proportional to its speed. When being towed at a steady speed of 3 m/s, the tension in the horizontal tow line is 200 kN. Ignoring transmission losses how much power must be developed by the ship's engines for a speed of 10 m/s? If at this speed the engines are shut off, how far will the ship travel before coming to rest?

At 3 m/s, the resistance to motion equals the tension in the tow line, hence

$$200\,000 = 3k$$

and therefore the constant of proportionality

$$k = 200\,000/3 \text{ Ns/m}$$

At 10 m/s

$$P = \text{rate of doing work against the resistance} = (k \times 10) \times 10 \text{ W}$$

$$= 6.67 \text{ MW}$$

While slowing down with the engines shut off the equation of motion gives

$$kv = ma = mv \, dv/ds$$

therefore

$$\int_0^s ds = \frac{m}{k} \int_0^{10} dv$$

$$s = \frac{20 \times 10^6 \times 3 \times 10}{200\,000} = 3000 \text{ m}$$

Example 5.6

An electric motor delivers 8 kW over the speed range 500 to 1500 rev/min. It is used to drive a fan which has a reisstance $R = (3.66 + 0.0038\omega^2)$ Nm where ω is the speed in rad/s. At what steady speed will the system run?

$$P = T\omega = 8000 \text{ W} = (3.66 + 0.0038\omega^2) \, \omega$$

hence

$$3.66\omega + 0.0038\omega^3 = 8000$$

By trial solution

$$\omega = 125.7 \text{ rad/s} = 1200 \text{ rev/min}$$

5.6   EFFICIENCY

It was noted earlier that the conversion of work to energy and subsequent recovery always involves some losses. The work done against friction is a common source of such losses, and has already been illustrated in some of the worked examples. However, there are other sources of energy loss, such as oil churning in transmission systems and wind resistance in most machines, which must sometimes be taken into account. An analytical treatment of these various energy losses is often made difficult through lack of data, but in appropriate cases the overall effect can generally be measured, or estimated from previous experience, and expressed in terms of an *efficiency*.

The efficiency of a machine is defined as the ratio

$$\eta = \frac{\text{useful work out}}{\text{total work in}}$$

and since the work terms are determined for the same time interval, efficiency may also be defined as

$$\frac{\text{power out}}{\text{power in}}$$

and is normally expressed as a percentage. When two or more machines having individual efficiencies $\eta_1, \eta_2, \eta_3, \ldots$ are connected in series, the overall efficiency is obtained by the product rule

$$\eta = \eta_1 \times \eta_2 \times \eta_3 \times \ldots \tag{5.11}$$

Example 5.7

An electric motor having an efficiency of 85 per cent drives a pump through a reduction gearbox having an efficiency of 80 per cent. The pump draws water from a sump and delivers it to a storage tank 20 m above through a 25 mm diameter pipe at the rate of 300 kg/min. Determine the electrical power input to the motor and the overall efficiency of the system. Losses in the pump may be assumed negligible, and the density of water may be taken as 1000 kg/m^3.

$$\text{Power out of pump} = \text{rate of gain of energy by water}$$

$$= \tfrac{1}{2}\text{mass flow rate} \times (\text{velocity})^2$$

$$+ \text{mass flow rate} \times g \times \text{height raised}$$

$$= \frac{1}{2} \times \frac{300}{60} \left(\frac{300 \times 4}{60 \times 1000 \times \pi \times 0.025^2} \right)^2$$

$$+ \frac{300}{60} \times 9.81 \times 20$$

$$= 259 + 981 \ \text{W}$$

$$= 1.24 \ \text{kW}$$

But

(motor input power) × 0.85 × 0.8 = 1.24 W

therefore

$P = 1.82$ kW

The *useful* work out is the potential energy gained so that the overall efficiency is

$\eta = 0.981/1.82 = 0.54$ or 54 per cent

SUMMARY

Work done on a system

by a force $\quad = \displaystyle\int F \cos \theta \, ds$

by a couple $\quad = \displaystyle\int M \, d\theta$

Gain of energy

potential PE $= mg(h_2 - h_1)$

strain SE $\quad = \frac{1}{2}k(x_2^2 - x_1^2)$

kinetic KE $\quad = \frac{1}{2}m(v_2^2 - v_1^2)$

Kinetic energy of a rigid body $= \frac{1}{2}mv^2 + \frac{1}{2}I_G\omega_2$

Work–Energy equation

Work done on a system = Gain of $(PE + SE + KE) +$ losses

Power

$P =$ Rate of doing work = Rate of gain of energy

Efficiency

$\eta = \dfrac{\text{Useful work out}}{\text{Total work in}} = \dfrac{\text{Power out}}{\text{Power in}}$

5.7   PROBLEMS

1. Compare the kinetic energies of (a) a 200 000 tonne (1 tonne = 1000 kg) tanker steaming at 20 km/h; (b) a 1500 kg motor car travelling at 80 km/h; (c) a 25 g bullet fired with a muzzle velocity of 950 m/s.
   [3.09 MJ, 370 kJ, 22.6 kJ]

2. How much useful energy has been expended getting a 9500 kg aircraft to an altitude of 4000 m and a speed of 500 km/h? Assume g is constant. [464.4 MJ]

3. During an emergency stop a motorist makes skid marks 14.8 m long, all four wheels being locked, on a surface where the coefficient of friction is 0.7. He claims that prior to applying the brakes he was not exceeding the speed limit of 50 km/h. Does the evidence support the claim? [No, $V = 51.3$ km/h]

4. In figure 5.7 the uniform 480 kg door can swing about a horizontal axis through O. The door is shown in its equilibrium position under the control of a spring AB of stiffness 5 N/mm. How much work must be done to pull the door down to its closed position where it makes an angle of 90° to the vertical? If the door is released from its closed and latched position, what will be its angular velocity as it passes through the equilibrium position?

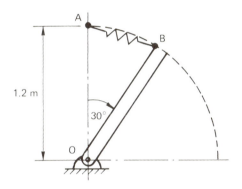

Figure 5.7

[1760 Nm, 3.91 rad/s]

5. When a gas is compressed rapidly in a cylinder, the pressure p and volume V of the gas are related by the law pV^γ is a constant. How much work must be done to compress the gas from an initial volume V_0 to a final volume V?
$$[\{1/V^{\gamma-1} - 1/V_0^{\gamma-1}\}\{p_0 V_0^\gamma/(\gamma - 1)\}]$$

6. A four-wheeled car has a total mass of 980 kg. Each wheel has a moment of inertia about its respective axle of 0.42 kgm² and a rolling radius of 0.32 m. The rotating and reciprocating parts of the engine and transmission are equivalent to a moment of inertia of 2.4 kgm² rotating at five times the road wheel speed. Calculate the total kinetic energy of the car when it is travelling at 90 km/h. If the car brakes to a stop with a deceleration of 0.5g, what is the mean rate of heat generation, assuming that the engine stalls just as the vehicle comes to rest? [494 kJ, 96.9 kW]

7. Within certain speed limits the drag D on an aircraft has the form $D = aV^2 + b/V^2$ where a and b are constants. Derive expressions for the minimum drag, and the minimum power required P_{min} for cruising at a steady speed in level flight, and the speeds at which these occur. Assuming constant efficiency, would either of these two conditions give the minimum fuel consumption for a given distance? If the maximum power (in terms of thrust) P_{max} is constant, what is the maximum climb rate for an aircraft of mass m?

   $[2\sqrt{(ab)}$ at $v^4 = b/a,\ 4(ab^3/27)$ at $v^4 = b/3a,\ (P_{max} - P_{min})/mg]$

8. In a pilot scheme for a power generation system, a barrage is built across a tidal basin. The basin has an area of 12 hectares (1 hectare $= 10\,000\ \text{m}^2$) and the sides may be assumed vertical over the tidal drop. The density of sea water is $1030\ \text{kg/m}^2$. The mean sea level in the basin falls in a sinusoidal manner through a total height of 3.5 m in a period of 6 hours, passing out to the sea through water turbine and generator sets in the barrage. During this period 7.3 GJ of electrical energy are generated. What is the overall efficiency of conversion of potential to electrical energy? Calculate also the mean and maximum power output, assuming the conversion efficiency is constant.

   [54 per cent, 338 kW, 531 kW]

6 Impulse and Momentum

6.1 LINEAR IMPULSE AND MOMENTUM

In chapter 2, Newton's second law was interpreted to give equation 2.2, repeated below

$$\Sigma F = R = ma$$

But in equation 3.2 the acceleration of a particle was defined as

$$a = dv/dt$$

so that

$$\Sigma F = R = m\,dv/dt$$

hence

$$\int_{t_1}^{t_2} R\,dt = m \int_{v_1}^{v_2} dv = m(v_2 - v_1) \tag{6.1}$$

The product of mass and velocity is defined as *linear momentum*, and the right-hand side of equation 6.1 is thus seen to be equal to the gain of linear momentum over the interval $t_2 - t_1$.

The left-hand side of equation 6.1 is defined as *linear impulse*. Note that the impulse term must include all forces acting on the particle, whereas in the previous chapter *work* involved only *active* forces, that is, those whose points of application have components in the direction of the respective displacements.

Example 6.1

A 5400 kg spacecraft is required to make a mid-course correction by increasing its speed by 50 km/h using its 4500 N thruster engine. How long should the burn last?

Using equation 6.1

$$\int 4500 \, dt = 5400(50/3.6)$$

$$t = 5400(50/3.6)/4500 = 16.7 \text{ s}$$

Impulse and momentum are both vector quantities so that equation 6.1 can be applied independently in two mutually perpendicular directions.

Example 6.2

The pressure in the barrel of a gun firing rises uniformly from zero to 4×10^8 N/m² over a period of 0.02 s, and then drops immediately back to zero as the shell leaves the end of the barrel. If the shell has a mass of 12 kg and a diameter of 65 mm, calculate the muzzle velocity. Determine also the efficiency of the gun if the chemical energy in the charge is 30 MJ. For the same charge and shell, what would be the effect on the muzzle velocity of shortening the barrel by 10 per cent?

The forces acting on the shell (taken as a particle) are shown in figure 6.1a. Applying equation 6.1 parallel and perpendicular to the barrel

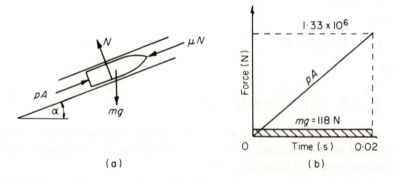

(a)　　　　　　　　　(b)

Figure 6.1

$$\int_0^{0.02} (pA - \mu N - mg \sin \alpha) \, dt = m(v - 0)$$

$$\int_0^{0.02} (N - mg \cos \alpha) \, dt = 0$$

Figure 6.1b shows the weight of the shell and the force on it due to the pressure in the barrel, as a function of time, and it is clear that in this case the integral of the former, that is, the area under the appropriate line, is negligible. The same is true of the normal reaction N and the friction force

μN, so that the impulse–momentum equations approximate to

$$\int_0^{0.02} pA \, dt = mv$$

therefore

$$v = \frac{1}{2}\left(4 \times 10^8 \times \frac{\pi}{4} \times 0.065^2 \times 0.02 \right)\Big/ 12$$

$$= 1106 \text{ m/s}$$

Efficiency, $\eta =$ useful work out/input

$$= \tfrac{1}{2} \times 12 \times 1106^2/30 \times 10^6$$

$$= 24.5 \text{ per cent}$$

At some general position

$$\int_0^t pA \, dt = \int_0^v m \, dv$$

putting $p = kt$ and integrating

$$kAt^2/2 = mv = m \, dx/dt$$

therefore, integrating again

$$kAt^3/6 = mx$$

hence, eliminating t between the above expressions

$$x = \sqrt{\left(\frac{2m}{9kA} \right)} v^{3/2} = \text{constant} \times v^{3/2}$$

Differentiating

$$dx = \frac{3}{2} \times \text{constant} \times v^{1/2} \, dv$$

$$\frac{dv}{v} = \frac{2 \, dx}{3 \, x}$$

therefore a 10 per cent reduction in x (barrel length) gives a 6.67 per cent reduction in muzzle velocity.

6.1.1 Linear Impact

When relatively large forces act for short intervals of time, such as occur in *collisions* or *impacts*, the resulting impulses are frequently very much bigger than those due to other more steady forces. This was seen to be the case in example 6.2, and by ignoring these other smaller impulses, a very simple solution is obtained with little loss of accuracy.

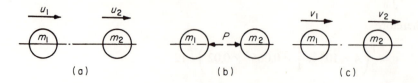

Figure 6.2   Linear impact (a) before impact; (b) during impact; (c) after impact

Figure 6.2 shows two particles, m_1 and m_2, just before, during, and just after impact along their line of centres. Assuming that the only significant impulse is that due to the reaction P between them, equation 6.1 can be applied to each particle to give

$$- \int P \, dt = m_1(v_1 - u_1)$$

$$\int P \, dt = m_2(v_2 - u_2)$$

hence, adding and re-arranging

$$m_1 u_1 + m_2 u_2 = m_1 v_1 + m_2 v_2 \tag{6.2}$$

Equation 6.2 shows that the combined linear momentum before impact is equal to the momentum after. This can be generalised to give the *principle of conservation of momentum* which states that *provided there are no external impulses acting on a system of particles, its momentum remains constant.*

During the impact shown in figure 6.2 some energy will generally be lost from the system due to the inelasticity of the particles, so that

$$\tfrac{1}{2} m_1 u_1^2 + \tfrac{1}{2} m_2 u_2^2 \geqslant \tfrac{1}{2} m_1 v_1^2 + \tfrac{1}{2} m_2 v_2^2$$

therefore

$$m_1(u_1 - v_1)(u_1 + v_1) \geqslant m_2(v_2 - u_2)(v_2 + u_2)$$

but from equation 6.2

$$m_1(u_1 - v_1) = m_2(v_2 - u_2)$$

hence

$$(v_2 - v_1) \leqslant -(u_2 - u_1)$$

It is usual to define a *coefficient of restitution*, e, so that

$$(v_2 - v_1) = -e(u_2 - u_1) \tag{6.3}$$

For the special case where no energy is lost on impact, e is 1, and this defines a *perfectly elastic impact*. More generally, the energy lost is

$$\Delta E = (\tfrac{1}{2}m_1 u_1^2 + \tfrac{1}{2}m_2 u_2^2) - (\tfrac{1}{2}m_1 v_1^2 + \tfrac{1}{2}m_2 v_2^2)$$

Using equations 6.2 and 6.3 to eliminate the velocities v_1 and v_2, this can be reduced to

$$\Delta E = \frac{1}{2}\left(\frac{m_1 m_2}{m_1 + m_2}\right)(u_1 - u_2)^2(1 - e^2) \tag{6.4}$$

From this it is clear that the maximum loss of energy during impact will occur when e is zero, and this defines a *plastic* or *inelastic impact*. Putting e equal to zero in equation 6.3, it is seen that after a plastic impact the two particles move off with the same velocity.

Experimental values of the coefficient of restitution vary widely, depending on shape and size, and tend to decrease as the impact velocity increases.

Example 6.3

An empty railway truck of mass 10 000 kg rolls at 2 m/s along a level track into a loaded truck of mass 20 000 kg which is stationary. If the trucks are automatically coupled together during impact, what is their common velocity? How much kinetic energy has been lost?

What velocity should the loaded truck have prior to the impact if they are both brought to rest after coupling?

From equation 6.2

$$m_1 u_1 = (m_1 + m_2)v_2$$

hence

$$v_2 = 10\,000 \times 2/(10\,000 + 20\,000) = 0.667 \text{ m/s}$$

The loss of kinetic energy is

$$\tfrac{1}{2}m_1 u_1 - \tfrac{1}{2}(m_1 + m_2)v_2^2$$
$$= \tfrac{1}{2}(10\,000) \times 2^2 - \tfrac{1}{2}(10\,000 + 20\,000) \times 0.667^2 = 13.3 \text{ kJ}$$

If the loaded truck has a velocity u_2 before impact such that the combined velocity afterwards is zero, then

$$m_1 u_1 + m_2 u_2 = 0$$

hence

$$u_2 = -10\,000 \times 2/20\,000 = -1 \text{ m/s}$$

The negative sign indicates that the loaded truck is travelling in the opposite direction.

Example 6.4

A pile-driver having a mass of 250 kg is dropped from a height of 2.3 m onto the top of a pile having a mass of 1300 kg. The ground resistance to penetration by the pile is 50 kN and the coefficient of restitution between the driver and the pile is 0.8. Determine the depth of penetration of the pile, the height to which the driver rises after impact, and the energy lost during impact. Find also the efficiency of the system, and the mean power that has to be supplied if there are 20 blows per minute.

By work–energy, the velocity of the driver just before impact is given by

$$0 = \tfrac{1}{2} \times 250 \times u_d^2 - 250 \times 9.81 \times 2.3$$

so that

$$u_d = 6.72 \text{ m/s}$$

From equation 6.3, the velocities after impact are

$$(v_p - v_d) = -0.8(0 - 6.72) = 5.37 \text{ m/s}$$

From the momentum equation, 6.2

$$250 \times 6.72 + 0 = 250v_d + 1300v_p$$

giving

$$v_p = 1.95 \text{ m/s} \qquad v_d = -3.42 \text{ m/s}$$

The minus sign for the velocity of the driver shows that it is a *rebound*, that is, upwards.

Using the work–energy equation for the pile after impact

$$-50\,000x = -1300 \times 9.81x - \tfrac{1}{2} \times 1300 \times 1.95^2$$

so that the depth of penetration x is

$$x = 0.0664 \text{ m}$$

For the driver

$$\tfrac{1}{2} \times 250 \times (-3.42)^2 = 250 \times 9.81 \, h'$$

therefore the driver rises to a height

$$h' = 0.596 \text{ m}$$

From equation 6.4, the energy lost during impact is

$$\Delta E = \frac{1}{2} \left(\frac{250 \times 1300}{1550} \right) (6.72 - 0)^2 (1 - 0.8^2)$$

$$= 1704 \text{ J}$$

The *useful* work per blow is that done against the ground resistance, and to maintain the operation, the driver must be raised back to its initial height after each blow. The efficiency is therefore

$$\eta = \frac{50\,000 \times 0.0664}{250 \times 9.81 \times 2.3} = 0.589 \text{ or } 58.9 \text{ per cent}$$

It should be noted that some of the useful work done is obtained from the potential energy lost by the pile, but since this is 'free' in the sense that it does not have to be 'paid for', it is reasonable to leave it out of the expression for efficiency. Also it is assumed that the kinetic energy of the driver due to the rebound velocity is of no value.

Power = rate of doing work

$$= 250 \times 9.81 \times 2.3 \times 20/60 \text{ W}$$

$$= 1.88 \text{ kW}$$

6.1.2   Oblique Impact

As already stated, impulse and momentum are both vector quantities, and thus equation 6.2 can be applied independently in two mutually perpendicular directions. For oblique impact problems, these directions are usually best taken along and perpendicular to the impact force. Equation 6.3 should be used with velocity components perpendicular to the contacting surfaces.

Example 6.5

In a game of snooker the cue ball is struck so that it hits a stationary red ball as shown in figure 6.3a. Ignoring friction effects, determine the velocities of the two balls immediately after impact if the coefficient of restitution is 0.9.

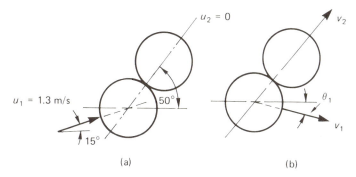

(a)   (b)

Figure 6.3   *(a) Just before impact; (b) just after impact*

Since there is no friction, the contact force must act along the line of centres and this is the direction in which the red ball will move off with velocity v_2, figure 6.3b. Applying equation 6.2 along and perpendicular to the line of centres, and cancelling the equal unknown masses

$$1.3\cos(50° - 15°) = v_1\cos(50° + \theta_1) + v_2$$
$$1.3\sin(50° - 15°) = v_1\sin(50° + \theta_1)$$

Applying equation 6.3 along the line of centres

$$v_2 - v_1\cos(50° + \theta_1) = -0.9[0 - 1.3\cos(50° - 15°)]$$

From the first and third equations

$$v_2 = \tfrac{1}{2} \times 1.3(1 + 0.9)\cos 35° = 1.012 \text{ m/s}$$

Substituting for v_2

$$v_1\cos(50° + \theta_1) = 1.3\cos 35° - 1.012 = 0.053 \text{ m/s}$$

Rearranging the second equation

$$v_1\sin(50° + \theta_1) = 1.3\sin 35° = 0.746 \text{ m/s}$$

hence

$$(50° + \theta_1) = \tan^{-1}[0.746/0.053] = 85.93°$$

so that

$$\theta_1 = 85.93 - 50 = 35.93°$$

and

$$v_1 = 0.746/\sin 85.93° = 0.748 \text{ m/s}$$

6.1.3 Recoil

The impulse–momentum principle can be applied to recoil problems in a very similar manner to that for impact described in the previous section. Initially, a body is moving with a certain velocity. It then breaks up into two or more pieces that separate relative to each other at their respective velocities. The process can be considered as an inverse inelastic collision; the total kinetic energy of the pieces will be greater than that of the original mass, and thus some energy must always be provided.

Example 6.6

A 40 kg missile is moving horizontally at an altitude of 75 m and at a speed of 230 m/s when it suddenly explodes into two equal masses. One of these falls vertically to the ground with zero initial velocity. How far down range will the other mass land? How much energy was released during the explosion?

After the explosion, one-half of the missile falls vertically to the ground 75 m below starting with zero velocity. The time taken can be found using equation 3.6

$$75 = 0 - \tfrac{1}{2}gt^2$$

hence

$$t = \sqrt{(2 \times 75/9.81)} = 3.91 \text{ s}$$

The other half will have an initial horizontal velocity v which can be found from the conservation of momentum principle

$$m \times 230 = (m/2)v$$

thus

$$v = 460 \text{ m/s}$$

In the vertical direction it will have the same motion as the first half and will thus continue at this horizontal speed for 3.91 s. The distance travelled down range will be

$$460 \times 3.91 = 1799 \text{ m}$$

The energy released will appear as an increase in kinetic energy

$$\tfrac{1}{2}(40/2) \times 460^2 - \tfrac{1}{2}40 \times 230^2 = 1.058 \text{ MJ}$$

6.1.4 Variable Mass

The concept of recoil can be extended to rocket propulsion and other variable mass problems. The only real difference is that the masses separate on a continuous rather than discrete basis.

Consider a rocket whose mass at some instant is m and which is travelling at a speed v. Thrust is generated by ejecting matter (exhaust gas) rearwards at a velocity v_r relative to the rocket. The absolute velocity of the exhaust gas is

$$v' = v - v_r$$

Applying the conservation of momentum principle over a time interval dt, during which the mass ejected is dm, leading to an increase in rocket speed dv, gives

$$mv = (m - \mathrm{d}m)(v + \mathrm{d}v) + (v - v_r)\,\mathrm{d}m$$

Rearranging and ignoring second-order terms

$$m\,\mathrm{d}v = -v_r\,\mathrm{d}m \tag{6.5}$$

and dividing by the time interval dt

$$m \, dv/dt = -v_r \, dm/dt \qquad (6.6)$$

Now dv/dt is the acceleration of the rocket and thus the left-hand side of equation 6.5 (mass × acceleration) is the resultant force on the rocket, that is, the thrust T. The right-hand side is the exhaust gas velocity relative to the rocket (generally constant for a given rocket motor) multiplied by the rate of change of rocket mass, and this last term is equal to the rate of fuel (and oxygen) consumption, with a change of sign. Thus, the thrust

$$T = \text{Exhaust gas relative velocity} \times \text{mass fuel rate} = v_r \dot{m}_f \qquad (6.7)$$

Example 6.7

The ratio of initial to 'all burnt' mass of a rocket is 4. What speed will the rocket reach when launched vertically from rest when all the fuel has been burnt with a constant relative exhaust gas velocity of 2400 m/s over a period of 60 s? Assume g constant.

From equation 6.6, with the addition of rocket weight

$$m \, dv/dt = -mg - v_r \, dm/dt$$

hence

$$\int_0^{60} g \, dt + \int_0^v dv = -v_r \int_{4m}^m (1/m) \, dm$$

so that

$$60g + v = v_r \log_e(4)$$

and

$$v = 2400 \log_e(4) - 60 \times 9.81 = 2738 \text{ m/s}$$

6.2   MOMENTUM OF A RIGID BODY

The linear momentum of a particle moving with a velocity v has already been defined as mv, and therefore the total linear momentum of the system of particles shown in figure 6.4 will be $\Sigma m_i v_i$. Putting

$$v_i = v_G + r_i(\dot{\theta})$$

$$\text{linear momentum} = v_G \Sigma m_i + \dot{\theta} \Sigma m_i r_i$$

$$= m v_G \qquad (6.8)$$

The moment of the linear momentum of the particle about some arbitrary point O is defined as the *angular momentum about O*, and by summing for all particles

$$\text{total angular momentum about O} = \Sigma m_i (\dot{y}_i x_i - \dot{x}_i y_i)$$

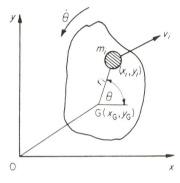

Figure 6.4   Angular momentum

Making the substitution

$$x_i = r_i \cos \theta + x_G$$
$$y_i = r_i \sin \theta + y_G$$

and expanding in a similar manner to that used in section 4.2

angular momentum about $O = m\dot{y}_G x_G - m\dot{x}_G y_G + I_G \dot{\theta}$ (6.9)

By integrating equation 4.7

$$\int \Sigma M_O \, dt = mx_G \int d\dot{y}_G - my_G \int d\dot{x}_G + I_G \int d\dot{\theta}$$

$$= mx_G(\dot{y}_{G2} - \dot{y}_{G1}) - my_G(\dot{x}_{G2} - \dot{x}_{G1}) + I_G(\dot{\theta}_2 - \dot{\theta}_1) \quad (6.10)$$

The left-hand side is defined as the sum of the *angular impulses about* O while the right-hand side is seen to be the gain in angular momentum about O.

Two special cases often arise. First, if the reference point is at G then equation 6.10 gives

$$\int \Sigma M_G \, dt = I_G(\dot{\theta}_2 - \dot{\theta}_1)$$ (6.11)

and second, if the reference point O is fixed, equation 6.10 gives

$$\int \Sigma M_O \, dt = I_O(\dot{\theta}_2 - \dot{\theta}_1)$$ (6.12)

which can also be derived directly by integration of equation 4.9.

Example 6.8

A door is arranged so that when it is swung open it strikes a rubber stop. How far from the hinge axis should the stop be positioned so that there is no impulsive reaction at the hinge?

Let the width of the door be a and the distance of the stop from the hinge axis be b. If the impulsive force on the stop during impact is P then application of equation 6.12 gives

$$\int Pb \, dt = (ma^2/3)(\omega_2 + \omega_1)$$

where ω_1 and ω_2 are the angular velocities of the door just before and just after impact. Note that this is a vector equation and due care must be taken with signs, thus, the angular velocity just before impact ω_1 will be in the opposite direction to the impulsive torque Pb and the rebound angular velocity ω_2.

In a similar manner equation 6.1 gives

$$\int P \, dt = m(v_2 + v_1)$$

where v_1 and v_2 are the linear velocities of the mass centre of the door just before and just after impact and the same comment about signs applies.

Putting $v_1 = r\omega_1$ and $v_2 = r\omega_2$ where $r = a/2$, eliminating the integral and simplifying

$$b = 2a/3$$

Example 6.9

A disc A having a moment of inertia of 4 kgm^2 and an angular velocity of 400 rev/min about a vertical axis is dropped onto a second disc B having a moment of inertia of 5 kgm^2 and an angular velocity of 200 rev/min in the opposite direction. Find the common velocity when all slipping has ceased, and the energy lost from the system. If the friction torque between the two discs is constant at 20 Nm, for how long does slip occur?

Applying equation 6.11 for each disc

$$-\int T \, dt = 4(n - 400) \times 2\pi/60$$

$$\int T \, dt = 5[n - (-200)] \times 2\pi/60$$

from which

$$n = 66.7 \text{ rev/min}$$

$$\Delta E = \left[\left(\frac{1}{2} \times 4 \times 400^2 + \frac{1}{2} \times 5 \times 200^2 \right) - \frac{1}{2} \times 9 \times 66.7^2 \right] \left(\frac{2\pi}{60} \right)^2$$

$$= 4390 \text{ Nm}$$

Since $T = 20$ Nm

$$-20t = 4(66.7 - 400) \times 2\pi/60$$

$$t = 7.00 \text{ s}$$

6.2.1 Impact of Rigid Bodies

When impact of a rigid body occurs there is generally only one point of impact and thus only one force giving rise to a significant impulse. If this point of impact is chosen as the point O in equation 6.10, it follows that the left-hand side must be zero and hence the angular momentum about O is conserved.

Example 6.10

A wheel of diameter 100 mm and radius of gyration 40 mm is rolling with a constant velocity along a horizontal surface when it meets a vertical step of height 10 mm. Assuming no slip or rebound, what should be the initial velocity of the wheel if it is just to mount the step?

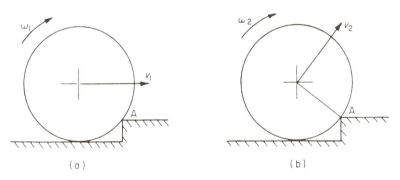

Figure 6.5 (*a*) *Just before impact*; (*b*) *just after impact*

Figure 6.5 shows the wheel just before and just after impact at A. Equating the angular momentum about A before and after impact

$$mv_1(0.05 - 0.01) + m \times 0.04^2\omega_1 = mv_2 \times 0.05 + m \times 0.04^2\omega_2$$

For no slip

$$v_1 = 0.05\omega_1 \qquad v_2 = 0.05\omega_2$$

therefore

$$v_1 = 1.14v_2$$

After impact, for the wheel just to mount the step, the work–energy principle gives

$$0 = mg \times 0.01 + \frac{1}{2}m\left[0 - \left(v_2^2 + \frac{0.04^2}{0.05^2}v_2^2\right)\right]$$

hence

$v_2 = 0.346$ m/s

and therefore

$v_1 = 0.394$ m/s

6.3 GYROSCOPIC EFFECTS

A conventional gyroscope consists of a symmetrical rotor spinning rapidly about its axis and free to rotate about one or more perpendicular axes. Figure 6.6 shows a completely *free gyroscope*. The rotor spins about its own

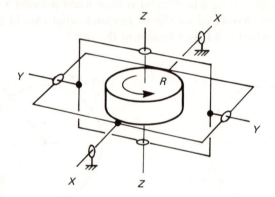

Figure 6.6

axis ZZ, and is gimbal mounted in an inner frame free to rotate about a perpendicular axis YY. This frame is, in turn, gimbal mounted in an outer frame free to rotate about the third axis XX, perpendicular to both YY and ZZ. Such a gyroscope exhibits three characteristics, namely *stability*, *nutation* and *precession*.

Stability   Ignoring gimbal friction, it is not possible to apply any torque to the rotor in figure 6.6 and hence the rotor axis ZZ remains pointing in a fixed direction in space. This, of course, applies whether the rotor is spinning or not, but as will be shown, the effect is greatly enhanced at high spin rates. Such devices form the basis of many navigation instruments.

Nutation   If, say, the outer frame is given an initial angular velocity about axis XX, then a periodic motion will ensue, made up of angular oscillations about axes XX and YY.

Precession   If the outer frame is subjected to a constant torque about axis XX then the inner frame will rotate about axis YY with a constant angular velocity.

The earth, which is a very massive free gyroscope, exhibits all three of these characteristics. Its polar axis remains almost fixed in the direction of the North star (Polaris) irrespective of its transit around the sun; owing to the earth's lack of sphericity the polar axis precesses, sweeping out a cone of apex angle nearly $47°$ in a period of $25\,800$ years; and there is a nutation arising from and superimposed on this such that the polar axis sweeps out a cone whose diameter at the North Pole is about 8 m, in 428 days.

When first encountered, the behaviour of a free gyroscope seems to raise doubts about the truth of Newton's laws, but in the next section these very laws will be used to explain one of the observed effects.

6.3.1   Rectangular Precession

Consider the particular case of a rotor of moment of inertia I spinning about its axis ZZ with an angular velocity ω, figure 6.7a. Using the right-hand

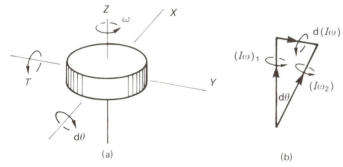

Figure 6.7

screw rule for vector notation (section 2.2.2) the angular momentum of the rotor can be represented by the vector $(I\omega)_1$, figure 6.7b. Now suppose that the whole assembly rotates through a small angle $d\theta$ about the x-axis, so that the new angular momentum is $(I\omega)_2$. Note that the angular momentum diagram is drawn in the y–z plane. The change of angular momentum $d(I\omega)$ will be the vector difference and is also shown in figure 6.7b, and will be

$$d(I\omega) = I\omega \, d\theta$$

Using the angular equivalent of Newton's second law, we get

Rate of change of angular momentum $= d(I\omega)/dt = I\omega \, d\theta/dt$

$$T = I\omega\Omega \qquad (6.13)$$

The left-hand side of equation 6.13 is the torque required to produce the precession velocity Ω about the x-axis. Examination of figure 6.7b indicates that this torque is represented by a vector having an angular sense about the y-axis as shown.

It should be noted that the theory described in this section is for *rectangular precession* only, that is, the spin, precession and torque axes are mutually perpendicular.

Example 6.11

A pair of wheels and axle from a railway carriage have a combined mass of 850 kg, a radius of gyration of 350 mm and a rolling radius of 650 mm. The rails on which they run are 1.44 m apart. An imperfection in a joint in one rail has the effect of raising the wheel on that side by 1.5 mm over a distance of 150 mm. What is the gyroscopic couple when passing the imperfection at 120 km/h and what is its effect on the axle?

Treating the wheel assembly as the rotor, its spin velocity is

$v/r = (120/3.6)/0.650 = 51.28$ rad/s

The imperfection leads to a precession velocity

$\Omega = $ vertical wheel velocity/rail spacing

$= [0.0015/(150 \times 3.6/120 \times 1000)]/1.44 = 0.231$ rad/s

Hence the gyroscopic couple is

$(850 \times 0.350^2) \times 51.28 \times 0.231 = 1161$ Nm

which must be provided by the axle supports in a 'steering' sense and which tends to make the rising wheel trail.

SUMMARY

Impulse–momentum principle

$$\int_{t_1}^{t_2} R\,dt = m(v_2 - v_1)$$

Conservation of momentum during particle impact

$$m_1 u_1 + m_2 u_2 = m_1 v_1 + m_2 v_2$$

Coefficient of restitution $v_2 - v_1 = -e(u_2 - u_1)$

Rocket propulsion Thrust $= v_r \dot{m}_f$

Impulse–momentum principle in rotation

$$\int_{t_1}^{t_2} \Sigma M_G \, dt = I_G(\dot\theta_2 - \dot\theta_1)$$

$$\int_{t_1}^{t_2} \Sigma M_O \, dt = I_O(\dot\theta_2 - \dot\theta_1), \qquad O \text{ fixed}$$

Gyroscopic couple—rectangular precession $= I\omega\Omega$

6.4 PROBLEMS

1. A 0.2 kg cricket ball moving at 25 m/s is struck by a batsman and moves off with a velocity of 40 m/s in the opposite direction. If the ball is in contact with the bat for 0.004 s, what is the average contact force? How much energy has the cricket ball gained?
   [3250 N, 97.5 J]

2. A car having a mass of 1200 kg is travelling at 5 m/s when it is struck in the rear by a truck having a mass of 8000 kg travelling at 8 m/s in the same direction. If the coefficient of restitution between the two vehicles is 0.3, what will be their velocities after the impact and how much energy will be dissipated?
   [8.39 m/s, 7.49 m/s, 4366 J]

3. When supplies are parachuted into a remote area, the maximum deceleration on striking the ground must be limited to $5g$ to prevent damage. What is the maximum vertical velocity with which supplies may strike the ground if the vertical reaction on impact takes the form shown in figure 6.8 (see next page)?
   [6.38 m/s]

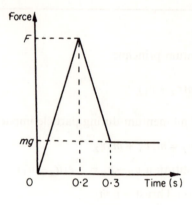

Figure 6.8

4. A 3000 kg truck is connected by a tow rope to an identical truck which is initially stationary. At the instant when the slack in the tow rope is just taken up, the towing truck has a speed of 2 m/s. Determine the common speed of the two trucks when the rope is stretched to its maximum, and the amount of energy that is stored in the rope at this instant. If the rope has a diameter of 30 mm, a modulus of elasticity of 192 kN/mm^2 and an ultimate tensile strength of 460 N/mm^2, what is the minimum length of rope necessary to prevent breakage? Assume the rope behaves elastically.
   [1 m/s, 3000 J, 7.70 m]

5. An automatic gun fires rounds having a mass of 0.1 kg with a muzzle velocity of 300 m/s. The moving parts of the breach have a mass of 5 kg and recoil against a spring of stiffness 20 N/mm. The spring is unstrained prior to firing. On the return stroke of the breach, a damping device is engaged that offers a constant resisting force. If only 20 per cent of the energy stored in the recoil spring is required to re-cock and re-load the gun, determine the force that should be exerted by the damper to bring the recoiling parts to rest in the firing position.
   [759 N]

6. When a helicopter is hovering, relatively large amounts of air are passed down through the rotors at a moderate velocity. When a VTOL (vertical take-off and landing) aircraft is hovering, relatively small amounts of exhaust gas are ejected downwards but at rather higher velocities. For the same total mass of helicopter and aircraft, use momentum and energy considerations to argue which of the two is likely to use less fuel in the hover. State carefully any important assumptions.

7. A ballistic pendulum consists of a 10 kg wooden block suspended by a light cable from a point 1.5 m above. When a 15 g bullet is fired into

the block it swings upwards to a maximum angle of 16°. What is the bullet velocity. Assume the dimensions of the block are small.
[712 m/s]

8. A ball is dropped from rest onto a horizontal surface. Show that the ratio of successive rebound heights is e^2 where e is the coefficient of restitution. What has happened to the momentum of the ball?

9. A ball strikes a horizontal surface at an angle θ and rebounds at an angle ϕ. What is the coefficient of restitution?
[$\tan \phi / \tan \theta$]

10. Two designs of a launch rocket for a satellite are under consideration. In the first design, the total launch mass is 12 000 kg of which 9000 kg is fuel and oxygen. In the second design two stages are used, comprising an 8000 kg first stage which includes 6000 kg of fuel and oxygen, and a 4000 kg second stage which includes 3000 kg of fuel and oxygen. When the first stage burn is complete, all unwanted parts are jettisoned without imparting any significant impulse to the second stage, which then continues alone. The exhaust gas velocities are the same for all rocket motors. Compare the final velocities achieved by the two designs. Note that both designs have the same launch mass and carry the same total amount of fuel. Ignore the effects of gravity.
[Two stage velocity:single stage velocity = 1.5:1]

11. In figure 6.9 the packing case is allowed to slide from rest down the 20° ramp where the coefficient of friction is 0.25. Assuming the leading edge is brought to rest at the bottom of the ramp, find the minimum value of x for the case to tip over into the vertical position.
[5.67 m]

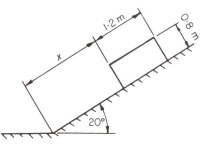

Figure 6.9

12. Figure 6.10a shows a schematic view of a free gyroscope used as an artificial horizon. Figure 6.10b–g shows the pilot's view of the instrument and the horizon through his cockpit window. Study the system and confirm that the behaviour is as shown.

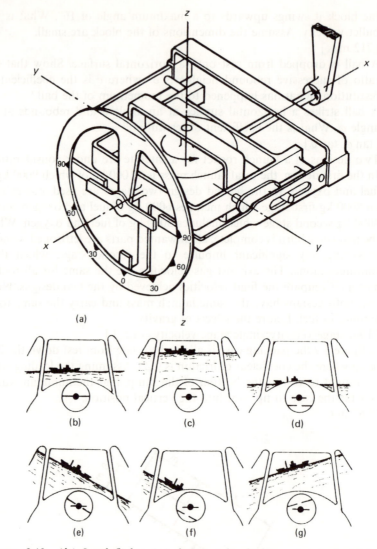

Figure 6.10 (b) Level flight; (c) diving; (d) climbing; (e) banked left turn; (f) climbing left turn; (g) diving right turn

13. A motor car travels round a left-hand bend of mean radius 25 m at a speed of 50 km/h. The rotating parts of the engine and transmission have an effective moment of inertia of 2.6 kgm² at a speed of 2000 rev/min clockwise when viewed from the front. What is the effect of the gyroscopic couple on the axle loads?
[Front increases and rear decreases by 144 N]

PART II: APPLICATIONS

7 Kinematics of Mechanisms

7.1 DEFINITION OF A MECHANISM

In chapter 1, a *machine* was described as a combination of bodies connected in such a way as to enable the transmission of forces and motion. A *mechanism* is a simplified model, frequently in the form of a line diagram, which will reproduce exactly the motion taking place in an actual machine. The various parts of the mechanism are called *links* or *elements*, and two links in contact and between which relative motion is possible, are known as a *pair*.

An arbitrary collection of links (forming a closed chain) that is capable of relative motion and that can be made into a rigid structure by the addition of a single link, is known as a *kinematic chain*. To form a mechanism from such a chain, one of the links must be fixed, but since any link can be chosen, there will be as many possible mechanisms as there are links in the chain. This technique of obtaining different mechanisms by fixing in turn the various links in the chain is known as *inversion*.

These concepts will now be considered in more detail.

7.1.1 Kinematic Pairs

The nature of the contact between the elements or links of a pair must be such as to give the required relative motion. The types of relative motion commonly called for are *sliding* such as occurs between a piston and cylinder; *turning* as with a wheel on an axle; and *screw motion* as between a nut and bolt. These three types form an exclusive class known as *lower pairs*. All other cases can be considered as combinations of sliding and rolling, and are known as *higher pairs*. Examples of these are meshing gear-teeth and cams.

7.1.2 Inversion

Consider the four-bar chain of figure 7.1 which consists of four links and four turning pairs. The link AD is shown as fixed and it can be seen that for

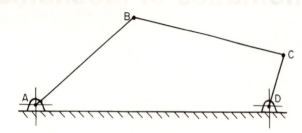

Figure 7.1   Four-bar chain

the lengths of links chosen, continuous rotation of link CD is possible, giving an oscillatory angular motion to link AB. Three other four-bar chains can be formed by inversion by fixing in turn the other three links, but continuous rotation of any one link may not always be possible. A common example of the four-bar chain is the double-wishbone independent front suspension for motor vehicles, with the wheel assembly and wishbones forming the three moving links, and the chassis acting as the fixed link, (see figure 2.24).

Figure 7.2a shows a slider–crank mechanism, consisting of four elements with three turning pairs and one sliding pair, which forms the basis of all reciprocating IC engines. By successive inversion three other mechanisms can be formed

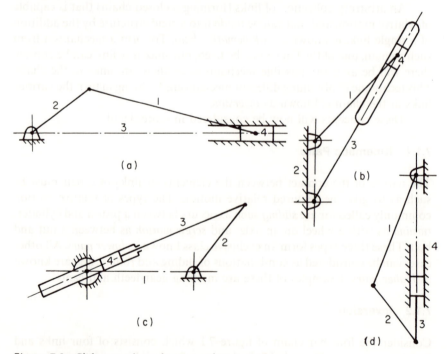

Figure 7.2   Slider–crank mechanism and its inversions

(a) figure 7.2b the Whitworth quick-return mechanism or Gnome aircraft engine
(b) figure 7.2c the oscillating-cylinder engine
(c) figure 7.2d a mechanism that is kinematically equivalent to the slider–crank but with the lengths of the crank and connecting-rod interchanged

One further example of inversion is shown in figure 7.3. Uniform rotation

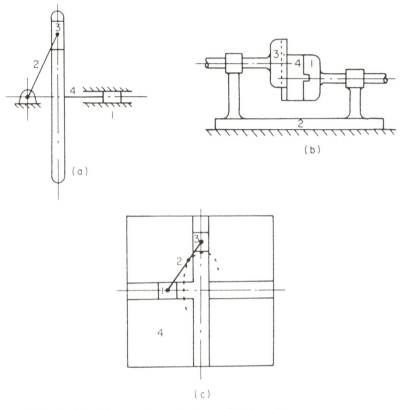

Figure 7.3   Double slider–crank mechanism and its inversions

of element 2 of the double-slider–crank mechanism or Scotch yoke (figure 7.3a) gives simple harmonic motion to the slider 4. By fixing element 2 (figure 7.3b) the device known as an Oldham coupling is formed, and this can be used to transmit motion between parallel shafts that are laterally displaced. In this case the slider 4 is in the form of a circular disc and has a diametral rib on each face engaging respective slots cut in discs on shafts 1 and 3. Fixing element 4 gives the elliptical trammel (figure 7.3c) while fixing element 3 leaves the mechanism unchanged.

7.3 DEGREES OF FREEDOM AND KINEMATIC CONSTRAINT

The position of a point that is free to move along a line can be specified by one co-ordinate and such a point is said to possess *one degree of freedom*. In two and three dimensions the point will have two and three degrees of freedom respectively since it will need a corresponding number of co-ordinates to specify its position. The co-ordinate system being used is immaterial but it is probably simplest to think in terms of cartesian co-ordinates, so that in addition to the three translatory degrees of freedom possessed by a point in space, a rigid body can rotate about each of the three axes giving a total of six degrees of freedom.

Any device that prevents one of these six possible movements of a rigid body is known as a *constraint*, and may be of two kinds

(a) *body closure* where movement is prevented by contact with an adjacent body, and
(b) *force closure* where movement is prevented by gravity or by a force from an elastic member.

In figure 7.4 the ball is free to roll over the horizontal surface and to spin about a vertical axis, thus having three rotational and two translational degrees of freedom. In the vertical direction, the ball is prevented from moving downwards by contact with the surface (body closure), but held against it by its own weight (force closure).

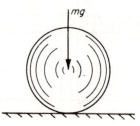

Figure 7.4   Force and body closure

If a body is to be completely fixed it will necessarily require six constraints, several of which may be by force closure. For *ideal* of *pure* kinematic constraint, the body closures should be single point contacts only, but clearly this condition can only be approached when loads are small, for example, with instrument mountings. If more than six constraints are applied some will be redundant and this condition should be avoided wherever possible.

In engineering practice, overconstraint must often be accepted. For example, to carry a reasonable load, bearing surfaces must be of finite area, giving many points of contact, and, while three are sufficient, a fourth leg may be added to a chair or table to give increased stability. A loaded rotating shaft may be supported in several bearings to prevent excessive vibration, even though it will result in some distortion in the shaft due to non-alignment of the bearings.

7.3   KINEMATICS OF MECHANISMS

In chapter 3 the kinematics of a particle and of a rigid body were considered in some detail. These principles must now be extended to a complete mechanism, but will be limited to two dimensions.

The techniques for finding the displacements, velocities and accelerations fall into one of two broad groups—graphical and analytical. The analytical technique has the great advantage of giving a general solution, that is, one that is valid for all configurations of the mechanisms, whereas graphical methods rely on scale drawings related to one particular configuration. A complete solution would thus require many drawings although modern computer methods can speed the process considerably. On the other hand, the amount of analysis and differentiation sometimes needed for apparently quite simple mechanisms may make the analytical method unmanageable. The choice depends very much on the nature of the mechanism and the tools available, and is largely a matter of experience.

7.3.1   Analytical Method

In this method the displacements of the elements of interest are defined with reference to some convenient datum, and then related geometrically. Successive differentiation will thus give expressions containing velocity and acceleration and hence the desired quantities can be obtained.

An important case where this method can be used with advantage is the slider–crank mechanism shown in figure 7.5. The crank OB rotates anticlockwise with a constant angular velocity ω. The displacement of the slider A measured from the crank axis O, is

$$x = r \cos \theta + l \cos \phi$$

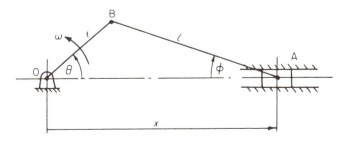

Figure 7.5

but

$$r \sin \theta = l \sin \phi$$

hence

$$x = r \cos \theta + 1 \sqrt{\left(1 - \frac{r^2}{l^2} \sin^2 \theta\right)}$$

The differentiation is made easier by first expanding the square-root term using the binomial theorem, so that

$$x = r \cos \theta + l\left(1 - \frac{r^2}{2l^2} \sin^2 \theta - \frac{r^4}{8l^4} \sin^4 \theta - \dots\right)$$

If the slider–crank is used as part of a reciprocating engine the ratio r/l is unlikely to exceed $\frac{1}{3}$ and since $\sin \theta \leqslant 1$, the successive terms in the expansion diminish rapidly. An acceptable approximation is thus obtained by taking only the first two terms, giving

$$x = r \cos \theta + l - \frac{r^2}{2l^2} \sin^2 \theta$$

Differentiating, and noting that $\omega = \dot{\theta}$ (in the same sense)

$$\dot{x} = -r\omega\left(\sin \theta + \frac{r}{2l} \sin 2\theta\right) \tag{7.1}$$

$$\ddot{x} = -r\omega^2\left(\cos \theta + \frac{r}{l} \cos 2\theta\right) \tag{7.2}$$

Example 7.1

For the slider–crank shown in figure 7.5, derive expressions for the components of the velocity and acceleration of a point P on the link AB that is distance b from A.

Putting x_P and y_P as the co-ordinates of the point P from an origin at the crank axis O

$$x_P = x - b \cos \phi$$

$$y_P = b \sin \phi$$

but

$$r \sin \theta = l \sin \phi$$

hence

$$x_P = x - b \sqrt{\left(1 - \frac{r^2}{l^2} \sin^2 \theta\right)} \approx x - b\left(1 - \frac{r^2}{2l^2} \sin^2 \theta\right)$$

$$y_P = \frac{rb}{l} \sin \theta$$

Differentiating and substituting for $\dot{x}$ from equation 7.1

$$\dot{x}_P = -r\omega \left[\sin \theta + \frac{r}{2l}\left(1 - \frac{b}{l}\right) \sin 2\theta \right]$$

Also

$$\dot{y}_P = r\omega \frac{b}{l} \cos \theta$$

Differentiating again

$$\ddot{x}_P = -r\omega^2 \left[\cos \theta + \frac{r}{l}\left(1 - \frac{b}{l}\right) \cos 2\theta \right]$$

$$\ddot{y}_P = -r\omega^2 \frac{b}{l} \sin \theta$$

Example 7.2

Figure 7.6a shows a Geneva mechanism whereby constant speed rotation of shaft O gives intermittent rotation of shaft P. The distance between shaft centres is 100 mm and the distance OA is 35 mm. For the position shown, find the angular velocity and acceleration of the slotted member.

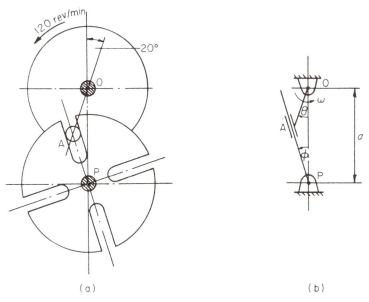

(a)   (b)

Figure 7.6   Geneva mechanism

Figure 7.6b shows a simplified model, and from the geometry of the mechanism

$$\tan \phi = r \sin \theta / (a - r \cos \theta)$$

Differentiating

$$\sec^2 \phi \, \dot{\phi} = \frac{(a - r \cos \theta) r \cos \theta \, \dot{\theta} - r^2 \sin^2 \theta \, \dot{\theta}}{(a - r \cos \theta)^2}$$

Rearranging and simplifying

$$\dot{\phi} = -\frac{r\omega(a \cos \theta - r)}{(a^2 + r^2 - 2ar \cos \theta)}$$

where $\omega = -\dot{\theta}$. Differentiating again and simplifying

$$\ddot{\phi} = -\frac{ar(r^2 - a^2)\omega^2 \sin \theta}{(a^2 + r^2 - 2ar \cos \theta)^2}$$

Substituting $a = 100$ mm, $r = 35$ mm, $\theta = 20°$ and $\omega = 120$ rev/min into the expressions for $\dot{\phi}$ and $\ddot{\phi}$ gives

$$\dot{\phi} = -5.40 \text{ rad/s} \qquad \ddot{\phi} = -75.6 \text{ rad/s}^2$$

both clockwise.

Example 7.3

Figure 7.7a shows a plan view of a camera C tracking an experimental car travelling at a constant speed along the line AA a distance a away. Sketch curves to show the angular velocity and acceleration of the camera as functions of its angular displacement θ. Indicate the salient values on the curves.

The displacement of the car relative to the datum normal to AA through the camera is

$$x = a \tan \theta$$

hence

$$\dot{x} = v = a \sec^2 \theta \, \dot{\theta}$$

and

$$\ddot{x} = 0 = a(\sec^2 \theta \, \ddot{\theta} + 2 \sec^2 \theta \tan \theta \, \dot{\theta}^2)$$

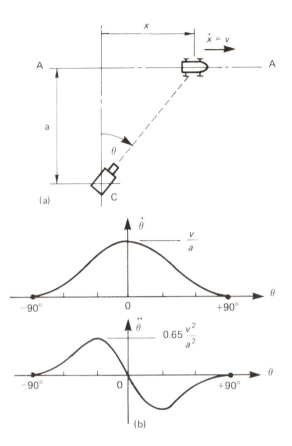

Figure 7.7

Rearranging

$$\dot\theta = \frac{v}{a}\cos^2\theta$$

and

$$\ddot\theta = -2\tan\theta\,\dot\theta^2 = -\frac{2v^2}{a^2}\sin\theta\cos^3\theta$$

The maximum angular velocity is clearly v/a and occurs when θ is zero. The maximum acceleration is not obvious and must be found by further differentiation *with respect to* θ. Thus

$$\frac{\mathrm{d}\ddot\theta}{\mathrm{d}\theta} = -\frac{2v^2}{a^2}(\cos^4\theta - 3\cos^2\theta\sin^2\theta)$$

and for this to be zero we have

$$\cos\theta = 0 \qquad \text{and} \qquad \cos^2\theta = 3\sin^2\theta$$

from which

$$\theta = \pm 90° \qquad \text{and} \qquad \pm 30°$$

The first two solutions give $\ddot{\theta} = 0$ while the second two give $\ddot{\theta} = \pm 0.650 v^2 / a^2$, and the curves are shown in figure 7.7b and c.

7.3.2 Instantaneous Centres

It should be clear from the previous example that the analytical method has severe limitations. The use of *instantaneous centres* avoids the complexity arising from the geometry of the mechanism and will often give a simple and elegant solution. The concept of the *instantaneous centre of zero velocity* has already been met in chapter 3, and is summarised for an arbitrary motion.

Consider a body in plane motion and suppose that the absolute velocities of two points A and B are as shown in figure 7.8a. If the velocities v_A and

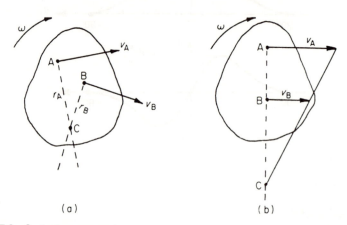

Figure 7.8   Instantaneous centres

v_B are different, then the body must have an angular velocity ω, and at the instant shown, point A must be rotating about a centre on a line through A perpendicular to v_A. In a similar manner, point B must be rotating about a centre on a line through B perpendicular to v_B. It therefore follows that the intersection of these two lines, C, must be the instantaneous centre of rotation. (It should be noted that it is only necessary to know the lines of action of v_A and v_B in order to locate C, and since in general the velocities v_A and v_B will vary from instant to instant, C is not a fixed point.) Having located the instantaneous centre, the relationship

$$\omega = v_A / r_A = v_B / r_B \tag{7.3}$$

immediately follows. Thus if say v_A is known completely and the line of action only of v_B is known, scale drawing and measurement of r_A and r_B will give the magnitude of v_B, and indeed the velocity of any other point on the body.

An important special case arises when the two velocities are parallel, figure 7.8b. C is then founded by the simple ratio

$$\omega = v_A/AC = v_B/BC$$

but the magnitudes of *both* velocities must be known. If v_A and v_B are opposite in direction, then C will lie between A and B.

Example 7.4

A slider–crank mechanism has a crank radius of 160 mm, and a connecting link 500 m long. The crank rotates at a steady speed of 4500 rev/min. Find the velocity of a point on the connecting link 360 mm from the slider, when the crank has turned through an angle of 40° measured from the position where the slider is furthest from the crank axis (outer dead-centre position). Compare the answer with that obtained by substitution of numerical values into the expressions derived in example 7.1.

Figure 7.9 shows a scaled drawing with the construction lines for finding

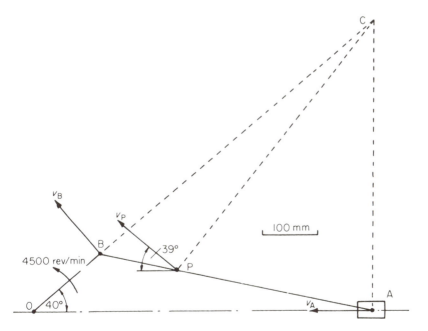

Figure 7.9

the instantaneous centre C of the connecting link. From equation 7.3

$$v_P = OB\omega_{OB} \times \frac{PC}{BC}$$

$$= 0.16\left(4500 \times \frac{2\pi}{60}\right) \times \frac{0.564}{0.640}$$

$$= 66.48 \text{ m/s}$$

and is in the direction shown.

From example 7.1

$$\dot{x}_P = -0.16\left(4500 \times \frac{2\pi}{60}\right)\left[\sin 40° + 0.16\left(1 - \frac{0.36}{0.5}\right)\sin 80°\right]$$

$$= -51.79 \text{ m/s}$$

and

$$\dot{y}_P = 0.16\left(4500 \times \frac{2\pi}{60}\right)\left(\frac{0.36}{0.5}\right)\cos 40°$$

$$= 41.59 \text{ m/s}$$

hence

$$v_P = 66.48 \text{ m/s}$$

at an angle $\tan^{-1}(41.59/51.79) = 39°$.

Example 7.5

A helicopter flies horizontally forwards at 82 km/h while its rotor turns at 150 rev/min clockwise when viewed from above. Assuming the blades rotate in a horizontal plane, locate their instantaneous centre.

A plan view of the rotor system is shown in figure 7.10. The tip velocities

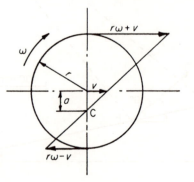

Figure 7.10

of two blades which are perpendicular to the velocity of the helicopter can be found by the principle of relative velocity. The instantaneous centre is at C and by similar triangles

$$\frac{v + r\omega}{a + r} = \frac{r\omega - v}{r - a}$$

hence

$$a = \frac{v}{w} = \frac{82}{3.6} \times \frac{60}{150 \times 2\pi} = 1.45 \text{ m}$$

As defined, the instantaneous centre relates to absolute velocities but it is perfectly proper to consider the instantaneous centre of one body *relative to another*. For example, in figure 7.9 the instantaneous centre of the connecting rod AB relative to a stationary reference is at C, but the instantaneous centre of AB relative to the crank OB is at B. This concept can be usefully exploited through the *three centres in line theorem*.

Consider three links 1, 2 and 3 connected as shown in figure 7.11. It is

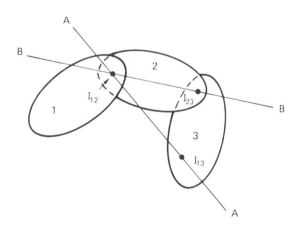

Figure 7.11

obvious that the instantaneous centre of link 1 relative to link 2 will be at I_{12}, that is, at the actual connecting pin. In a similar manner, the instantaneous centre of link 2 relative to link 3 will be I_{23}. At this stage the location of the instantaneous centre of link 1 relative to link 3 is unknown, but suppose it is at I_{13}. By definition of the instantaneous centre, *at the instant shown*, all points on link 1 are moving relative to link 3 as if they were rotating about I_{13}. In particular, I_{12} must be moving in a direction perpendicular to line AA. Using a similar argument, I_{12} must be moving in a direction perpendicular to line BB, but this is clearly impossible unless lines AA and BB are collinear, which means that I_{12}, I_{23} and I_{31} must also be collinear. The three centres in line theorem is true for any three links, whether or not there are any direct connections between them (as shown in figure 7.11), but in most applications there are such connections. It should also be noted that I_{ij} is the same point as I_{ji}.

Example 7.6

The suspension unit shown previously in chapter 2 is repeated in figure 7.12 with the addition of the vertical axis of symmetry through the body. Locate the roll centre (the instantaneous centre of the body relative to the ground) for the position shown.

The various members are numbered 1 to 5, and the roll centre will be the location of I_{15}. The locations of the instantaneous centres I_{12}, I_{13}, I_{24} and I_{34} are obvious. The location of I_{45} assumes that during roll the wheel pivots about the centre of its contact area. I_{14} can be found by noting that it lies at the intersection of the lines through the two pairs of points I_{12} and I_{14}, and I_{13} and I_{14}. The roll centre I_{15} will lie on the lie through I_{14} and I_{45}, and, by symmetry, it must also lie on the centre line.

The location of roll centre is an important parameter in the study of vehicle dynamics. In figure 7.12 the suspension geometry was chosen to give

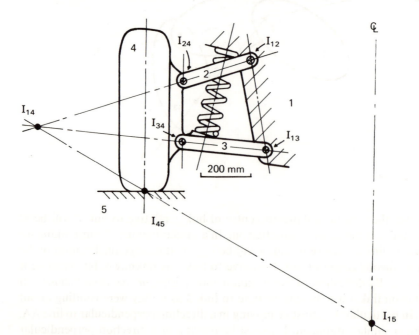

Figure 7.12

an open construction which is easy to follow, leading to an unrealistically low roll centre. A roll centre closer to the mass centre will minimise roll angles during cornering.

7.3.3 Velocity Diagrams

Although the method of instantaneous centres can be extended to more complex mechanisms, the use of *velocity diagrams* is generally simpler. The technique uses the concept of relative velocities, working progressively through the mechanisms from a link whose velocity is initially known. Scale drawings are again necessary and thus many drawings will be required if a complete picture of the motion of the mechanism is to be built up. The widespread use of velocity diagrams has led to the adoption of a special notation, and this will be considered in isolation first.

Figure 7.13a shows a link in plane motion, the *absolute velocities* of two points A and B being v_A and v_B respectively. The *velocity of B relative to A*,

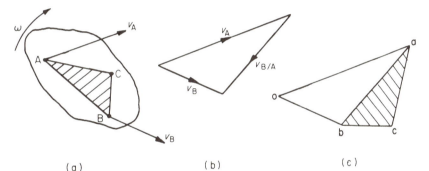

(a)   **(b)**   **(c)**

Figure 7.13   Notation for velocity diagrams (a) space diagram; (b) velocity vectors; (c) velocity diagram

$v_{B/A}$, can be determined by vector subtraction, as explained in section 1.3.1 and shown in figure 7.13b. Since AB is a fixed length, $v_{B/A}$ will of course be perpendicular to AB and of magnitude ABω. The conventional notation is to represent the velocity v_A (which might be more correctly written as $v_{A/O}$ where O is a fixed reference point), by a vector **oa**, where the order of the letters gives the sense, rather than using an arrowhead. In a similar manner v_B is represented by the vector **ob**. Hence $v_{B/A}$ is given by the vector **ab**, and because there are no arrowheads on the diagram, the same line taken in the opposite direction, that is, the vector **ba**, will represent $v_{A/B}$. This is consistent with the statement $v_{B/A} = -v_{A/B}$.

Consider now a third point C. Working relative to points whose velocity is already known, $v_{C/A}$ will appear on the velocity diagram as a line through point **a** perpendicular to AC. Similarly, $v_{C/B}$ will appear as a line through point **b** perpendicular to BC, thus defining point c (figure 7.13c). Because of the construction, triangles ABC and **abc** are geometrically similar and the latter is known as a *velocity image*. In general, any number of points on the space diagram that bear a fixed relationship to each other, will appear as a geometrically similar image on the velocity diagram.

Example 7.7

Use the velocity-diagram method to solve example 7.4. Find also the angular velocity of the connecting rod.

First a space diagram is constructed, figure 7.14a. Given the angular velocity of the crank

$$v_{B/O} = 0.16 \times (4500 \times 2\pi/60)$$

$$= 75.40 \text{ m/s perpendicular to OB}$$

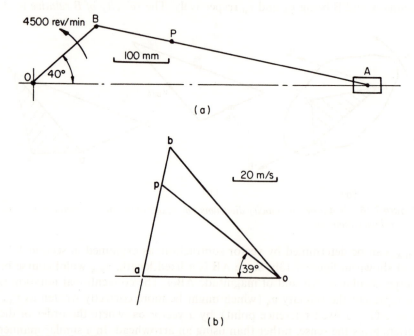

(a)

(b)

Figure 7.14   (a) Space diagram; (b) velocity diagram

This enables the vector **ob** to be drawn, figure 7.14b. Because of the constraints, v_A (that is, $v_{A/O}$) must be along the line of stroke, while $v_{A/B}$ will be perpendicular to AB, thus locating **a**. Since P is a fixed point on AB, then APB and **apb** will be similar (velocity image) and hence v_P can be determined in magnitude and direction, by measuring **op**. Hence

$$v_P = 66.48 \text{ m/s}$$

Also

$$v_{B/A} = \mathbf{ab} = AB \times \text{angular velocity of AB}$$

hence

$$\omega_{AB} = 58/0.5 = 116 \text{ rad/s clockwise}$$

In solving example 7.7, the velocity of one point on a link relative to another point on the same link was determined by using polar co-ordinates (section 3.1.3b) although its application was very simple, involving only the transverse component. When sliding occurs along an axis that is not fixed, the radial or sliding component must also be taken into account. Problems of this kind are generally made simpler by introducing two coincident points as shown in the following example.

Example 7.8

In the mechanism of figure 7.15a, D is the slider at the end of the link rotating about centre P, and C is a *fixed point* on AB coincident with D at the instant shown. OA = 0.6 m, AB = 1.9 m, BQ = 1.0 m and PD = 1.4 m. If the crank OA rotates anticlockwise at 120 rev/min, determine the angular velocity of the member PD, and the speed of sliding at D.

The construction of the velocity diagram, figure 7.15b, is as follows.

(a) $v_{A/O} = 0.6 \times 120 \times 2\pi/60 = 7.54$ m/s perpendicular to OA; draw **oa**.
(b) Since P and Q are fixed points, **o**, **p** and **q** will be coincident on the velocity diagram.
(c) $v_{B/Q}$ is perpendicular to BQ and $v_{B/A}$ is perpendicular to BA hence giving **b**.
(d) ACB is similar to **acb**.
(e) $v_{D/P}$ is perpendicular to DP and $v_{D/C}$ is along AB hence locating **d**.

From the completed diagram

$$v_{D/P} = \mathbf{pd} = 2.3 \text{ m/s}$$

therefore

$$\omega_{PD} = 2.3/1.4 = 1.64 \text{ rad/s}$$

Also, the sliding velocity is

$$v_{D/C} = \mathbf{cd} = 9.3 \text{ m/s}$$

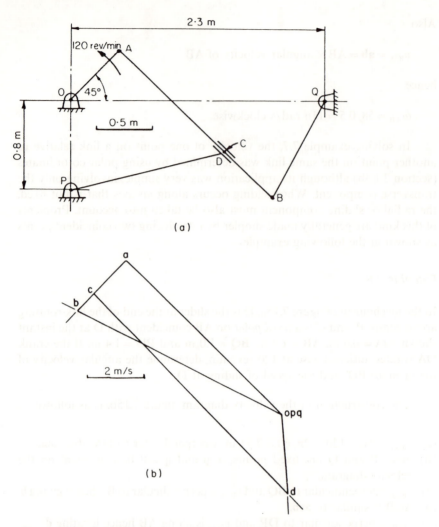

Figure 7.15   (a) Space diagram; (b) velocity diagram

7.3.4   Acceleration Diagrams

The use of polar co-ordinates for velocity diagrams will now be extended to acceleration diagrams, but care must be taken to include all the relevant terms (see equations 3.17 and 3.18). A notation similar to that described for velocity diagrams will be used but with a dash added to distinguish acceleration from velocity vectors.

Example 7.9

For the four-bar chain of figure 7.16a find the velocity and acceleration of point C, and the angular velocity and acceleration of link PB. AC = CB = 0.45 m, OA = 0.5 m, PB = 0.65 m and crank OA rotates clockwise with an angular velocity of 2.6 rad/s which is decreasing at 4.5 rad/s².

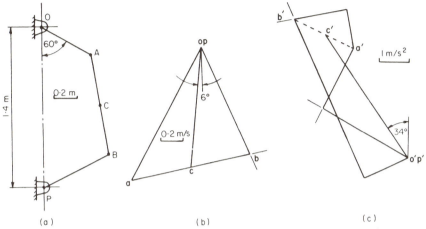

(a)   (b)   (c)

Figure 7.16   (a) Space diagram; (b) velocity diagram; (c) acceleration diagram

The steps in the solution are as follows.

(a) Draw the velocity diagram, figure 7.16b.

(b) $a_{A/O}$ consists of two terms

$$(a_{A/O})_r = 0.5 \times 2.6^2$$

$$= 3.38 \text{ m/s}^2 \text{ directed towards O}$$

and

$$(a_{A/O})_\theta = 0.5 \times 4.5$$

$$= 2.25 \text{ m/s}^2 \text{ perpendicular to OA}$$

in a sense consistent with the anticlockwise acceleration of OA. Starting from o′ these two components are now drawn to give point a′.

(c) $(a_{B/A})_r = 0.9 \times \omega_{AB}^2 = 0.9(\mathbf{ab}/0.9)^2$

$$= 1.30 \text{ m/s}^2 \text{ directed towards A}$$

$(a_{B/A})_\theta$ is perpendicular to AB but is unknown in magnitude at this stage.

(d) $(a_{B/P})_r = 0.65\omega_{PB}^2 = 0.65(\mathbf{pb}/0.65)^2$

$$= 160 \text{ m/s}^2 \text{ directed towards P}$$

$(a_{B/P})_\theta$ is perpendicular to PB but is also unknown in magnitude.

(e) The two known radial components from steps (c) and (d) are now drawn from their respective origins **a′** and **p′**, and the corresponding transverse terms drawn in direction only to locate **b′**.

(f) ACB will produce a geometrically similar *acceleration image* **a′c′b′** giving the line **o′c′**.

From the velocity diagram

$$v_{C/O} = \mathbf{oc} = 1.04 \text{ m/s}$$

$$\omega_{PB} = \mathbf{pb}/PB = 1.02/0.65 = 1.57 \text{ rad/s clockwise}$$

and from the acceleration diagram

$$a_{C/O} = \mathbf{o′c′} = 4.9 \text{ m/s}^2$$

$$\alpha_{PB} = (\text{component of } \mathbf{p′b′} \text{ perpendicular to PB})/PB$$

$$= 5.7/0.65 = 8.8 \text{ rad/s}^2 \text{ anticlockwise}$$

Example 7.10

Solve example 7.2 by drawing velocity and accceleration diagrams.

On the simplified space diagram, figure 7.17a, A is the slider on the end of the crank centre O, and B is a fixed point on the link centre P. The velocity diagram is first drawn in the usual way, figure 7.17b. The steps in the construction of the acceleration diagram, figure 7.17c are then

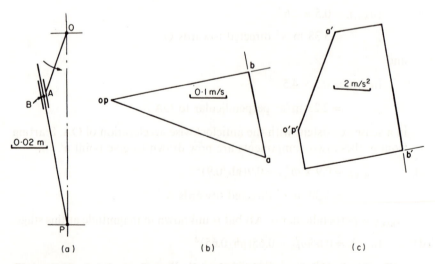

(a)   (b)   (c)

Figure 7.17 *(a) Space diagram; (b) velocity diagram; (c) acceleration diagram*

(a) $(a_{A/O})_r = 0.035 \times (120 \times 2\pi/60)^2 = 5.53 \text{ m/s}^2$

giving the vector $\mathbf{o'a'}$.

$(a_{A/O})_\theta = 0$ since ω_{OA} is constant.

(b) $(a_{B/P})_r = BP \times \omega_{PB}^2 = 0.068 \times (0.37/0.068)^2 = 2.01 \text{ m/s}^2$

$(a_{B/P})_\theta$ is known only in direction.

(c) Consider now the acceleration of B relative to A (the two coincident points). $(a_{B/A})_r$ is also known only in direction and in fact is the sliding acceleration ($\ddot{r}$-component of equation 3.17).

$(a_{B/A})_\theta = 2 \times \mathbf{ab} \times \omega_{PB} = 2 \times 0.23 \times (0.37/0.068) = 2.50 \text{ m/s}^2$

where $\mathbf{ab}$ is the sliding velocity. This is the Coriolis term, and is perpendicular to PB and in the direction consistent with ω_{PB}.

(d) Drawing the two known components from steps (c) and (d), and then the two components whose directions only are known, gives point $\mathbf{b'}$.

From the velocity diagram

$\omega_{PB} = \mathbf{pb}/PB = 0.37/0.068 = 5.40 \text{ rad/s clockwise}$

and from the acceleration diagram

$\alpha_{PB} = (\text{component of } \mathbf{p'b'} \text{ perpendicular to PB})/PB$

$= 5.2/0.068 = 76 \text{ rad/s}^2 \text{ clockwise}$

SUMMARY

Lower pairs: sliding, turning and screw

Higher pairs

Inversion of simple mechanisms

Degrees of freedom and constraint

Kinematics of mechanisms

Velocities and accelerations by differentiation

Instantaneous centres

Velocity diagrams

Acceleration diagrams

7.4   PROBLEMS

1. For the Scotch yoke of figure 7.3a show that rotation of the crank 2 at a constant speed will produce simple harmonic motion of the slider 4 at a frequency ω. Such a mechanism is used to compress refuse which gives a resisting force on the slider 4 proportional to its displacement. Ignoring losses and inertia effects, show that the power required to drive the system also varies harmonically, but at a frequency of 2ω.

2. Figure 7.18 shows a form of elliptical trammel (an inversion of the Scotch yoke) used to transmit motion between parallel offset shafts. Shaft O is rigidly attached to the midpoint of an arm AB, whose ends carry blocks that slide in mutually perpendicular slots in a disc attached to shaft P. If AO = OB = OP = a, determine the relationship between the angular velocities of shafts O and P. What is the maximum sliding velocity?
   $[\omega_O/\omega_P = \frac{1}{2}, a\omega_O]$

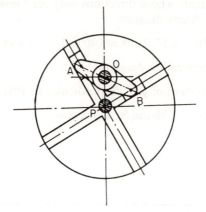

Figure 7.18

3. A slider–crank mechanism has a crank to connecting link ratio of $\frac{1}{4}$. If the diameter of the journal at the crank connecting-link bearing is 50 mm, determine the maximum rubbing velocity at the bearing when the crank rotates at 3800 rev/min.
   [12.4 m/s]

4. The 85 mm crank of a slider-crank mechanism has its axis of rotation offset from the vertical line of stroke by 50 mm to the left. The axis is below the slider and the connecting link is 500 mm long. Estimate the length of stroke and the ratio of times for the down-stroke to the up-stroke when the crank rotates uniformly in an anticlockwise direction. Use the method of instantaneous centres to find the velocity of the slider when the crank has turned through 230° measured from the position where the slider is uppermost, at a crank speed of 2000 rev/min.
   [0.978, 12.40 m/s]

5. For the mechanism of figure 7.19, use the method of instantaneous centres to find the velocity of point C and the angular velocity of link PB. OA = 600 mm, PB = 200 mm, AB = 820 mm, AC = 500 mm and BC = 400 mm.
[2.66 m/s, 20.4 rad/s anticlockwise]

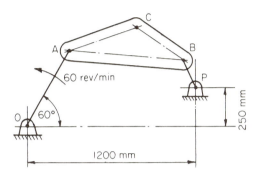

Figure 7.19

6. In figure 7.20, wheel A rotates anticlockwise at 200 rev/min about the fixed axis through O while the arm rotates clockwise at 100 rev/min about the same axis. The wheel B is free to rotate about an axis at the end of the arm, and engages wheel A without slip. Locate the instantaneous centre of wheel B and hence find its angular velocity.
[16.6 mm towards O from the centre of wheel B, 85 rad/s clockwise]

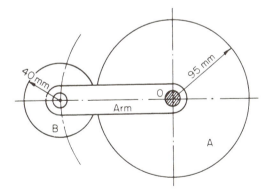

Figure 7.20

7. Figure 7.21 shows a strut type suspension. The upper telescopic link AB contains the spring element; its left-hand end A is rigidly attached to the wheel hub while the right-hand end B is pivoted to the vehicle body. The lower link CD is pivoted to the hub at C and to the body at D. Locate the roll centre for the body. The diagram is to scale, but the geometry has been distorted somewhat to give a more compact diagram for the solution.

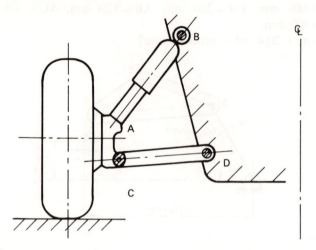

Figure 7.21

8. Draw velocity and acceleration diagrams for the slider–crank mechanism of problem 4 in the position given, and hence find the velocity and acceleration of the piston.
   [12.40 m/s, 2320 m/s^2]

9. Draw velocity and acceleration diagrams for the four-bar chain of problem 5, given that the link OA has an anticlockwise acceleration of 20 rad/s^2, and hence find the velocity and acceleration of point C, and the angular velocity and angular acceleration of PB.
   [2.66 m/s, 78 m/s^2, 20.4 rad/s anticlockwise, 455 rad/s^2 anticlockwise]

10. Figure 7.22 shows a hatch cover that can pivot about a horizontal axis through O, under the control of a ram AP. The ram is supplied with

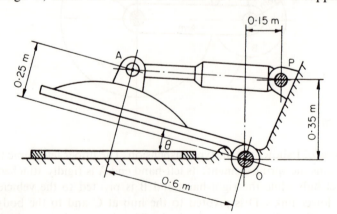

Figure 7.22

hydraulic fluid so that it extends at a constant rate of 50 mm/s. Draw velocity and acceleration diagrams for the hatch just as it closes, that is, when $\theta = 0$, and hence find the angular acceleration of the hatch and of the arm at this instant.

[0.010 rad/s^2, 0.0029 rad/s^2 both anticlockwise]

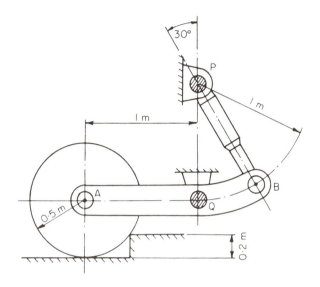

Figure 7.23

11. Figure 7.23 shows a trailing-arm active suspension for a vehicle. The arm AQB carries the road wheel at A and is pivoted to the chassis at Q. A ram is connected between the end of the arm B and a second point on the chassis P. In operation, the road profile is sensed and the resulting signal used to control the flow of fluid to the ram. This flow is such that the chassis continues to move in a horizontal straight line with a constant velocity of 10 m/s. For the position shown, where the wheel has just started to mount the step, determine the necessary velocity and acceleration of sliding between the piston and the cylinder wall of the arm.

[5.8 m/s extending, -337 m/s^2]

8 Friction and Lubrication in Machines

8.1 FRICTION, LUBRICATION AND WEAR

The nature of causes of friction, together with some of its commonly experience manifestations, have been discussed in chapter 2. Generally speaking the effect of friction in machines is either to prevent slipping or to dissipate energy between two contacting parts. For example, a vehicle needs a high friction force between the driving wheels and the road to achieve maximum acceleration and cornering speed without skidding, and a high brake friction to minimise stopping distance.

However, there is a penalty to pay for the presence of friction forces. When parts of a machine slide one over the other, as in any form of bearing, work done against friction is converted into heat, with a consequent loss of mechanical efficiency. At the same time, if direct material contact occurs, there will be surface wear of the mating parts. Over a period of time this will result in failure of the machine due to loss of material strength, excessive free play between members, or seizure of the bearing surfaces. It is interesting to note that, although wear is a consequence of friction, there is no theoretical relation between them. Wear rate may be greatly reduced by hard surfacing such as chromium plating without much affecting the coefficient of friction. Even untreated metals usually form an oxide coating, which tends to reduce friction but will accelerate or retard wear according to the relative hardness of the oxide and the parent metal. In general it has been found advantageous to use a hard metal sliding or rotating member in conjunction with a softer metal or plastic bearing. Steel shafts run well in bearings containing alloys of tin, lead, and copper, and more recently PTFE, nylon and other plastics have been successfully developed as bearing materials.

In this chapter those engineering applications that rely on friction for their functioning, such as clutches, belt drives and brakes, will be analysed. Also considered is the screw-and-nut type of problem, where friction results in loss of efficiency during tightening, but may be essential in preventing unloading of the thread. It is usual to assume that dry (unlubricated) conditions apply in these cases, the coefficient of friction being of the order of 0.2 to 0.4 for most pairs of materials. But first the influence of lubrication on friction properties will be discussed, and in particular its role in the reduction of bearing losses and wear.

Two distinct forms of lubrication can arise, depending mainly on speed and load values. The ideal condition is for the bearing surfaces to be separated by a film of lubricant, thus greatly reducing the frictional resistance and practically eliminating wear. For a given lubricant, film lubrication will be self-generating provided the ratio of speed:load in the bearing exceeds a minimum value and certain geometrical conditions are satisfied. Alternatively the film may be sustained by external pressurisation (as would be essential for low-speed applications). If, however, conditions do not permit a continuous film to be maintained, then only surface lubrication, known as boundary lubrication, can be achieved.

8.1.1 Boundary Lubrication

When the relative sliding speed is too low (or oscillating between zero and a maximum as in the case of the small end bearing of an engine connecting-rod) or the load intensity is too high, only boundary lubrication is possible. In these circumstances a thin layer of lubricant, perhaps only a few molecules in depth, may adhere to the surfaces to prevent direct contact. The tenacity with which it adheres to and is absorbed by the surfaces is a property of the lubricant known as 'oiliness'. Under optimum conditions a low value of coefficient of friction (usually between 0.05 and 0.10) is accompanied by good protection against seizure under high loads.

A number of materials and compounds have been found to possess good oiliness qualities. Lubricants are then made up by small additions of one of these to a mineral oil (partly to keep down cost, and also to provide a stable base). Among the additives used are the fatty acids and organic oils, sulphur and chlorine compounds, molybdenum disulphide, and graphite.

8.1.2 Film Lubrication

Under favourable conditions of speed and load, and provided the clearance space can adapt to a convergent wedge shape, a continuous film of lubricant will build up between the bearing surfaces as illustrated in figure 8.1 for journal and thrust bearings. This is referred to as *hydrodynamic* (film) lubrication, the bearing load being carried by the pressure generated in the film, as the lubricant is dragged into the convergent space by the motion of one surface relative to the other. Pressures up to about $7 \, \text{MN}/\text{m}^2$ are common and film thicknesses vary typically between 10^{-1} and 10^{-3} mm. The effective frictional force is then due only to the viscous resistance of the lubricant, and wear of the surfaces is completely eliminated (except for that which

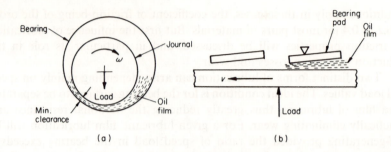

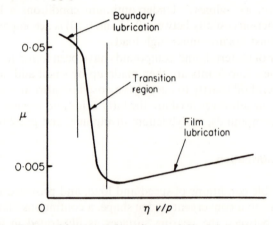

Figure 8.1   Hydrodynamic film lubrication (a) journal bearing; (b) tilting-rod thrust bearing

occurs under starting and stopping conditions, when boundary lubrication applies).

It can be shown that, in hydrodynamic lubrication, the effective coefficient of friction (defined as the ratio of total tangential resistance to load) is directly proportional to $\eta v/p$, where η is the viscosity of the lubricant, v is the relative sliding velocity of the bearing surfaces, and p is the nominal bearing pressure (ratio of load to projected area). This relation is represented graphically in figure 8.2, which indicates also the changeover from boundary to film lubrication as the velocity increases. There is a rapid fall of μ, by a factor of at least 10, through a transition region in which part boundary, part film conditions apply.

Figure 8.2   Coefficient of friction in boundary and film lubrication

8.2   FRICTION BETWEEN A SCREW AND NUT

When a torque (the 'effort') is applied to a nut in order to tighten it on a threaded bolt, the bolt is subjected to axial tension (the 'load'). Sliding friction forces are distributed over the thread surfaces in a complex and indeterminate manner, depending on the conformity of the thread profiles, the extent of pitch errors, and the elasticity of the materials. However, effort, load and friction forces are in a fixed relation (for given dimensions and materials), and consequently it is permissible to lump them together and to assume that they are in equilibrium at a point on the effective (that is, mean) radius of the screw. The force system is then equivalent to that for a mass, with a weight equal to the load, being pulled up an inclined plane (at the thread helix angle) against a frictional resistance. This forms the mathematical model on which the subsequent analysis is based.

8.2.1 Square-threaded Screw

Let W be the axial load and P the tangential effort required at the mean screw radius $\frac{1}{2}d$. If l is the lead of the thread (that is, pitch × number of starts), the helix angle of the thread is

$$\theta = \tan^{-1} l/\pi d \tag{8.1}$$

When there is relative movement the normal reaction N and friction μN may be combined into a reaction R which will be inclined at the friction angle ϕ to the normal. Figure 8.3 shows the system of forces in equilibrium, from which

$$P = W \tan(\theta + \phi) \tag{8.2}$$

Since this is the effort that would have to be applied at the screw radius, it corresponds to a torque T given by

$$T = P \times \tfrac{1}{2}d = \tfrac{1}{2}W d \tan(\theta + \phi) \tag{8.3}$$

(from equation 8.2).

In practice this torque may be applied through a gear reduction (as in a mechanical lifting-jack) or a simple lever system (as in a machine vice). Either of these arrangements would further reduce the effort required.

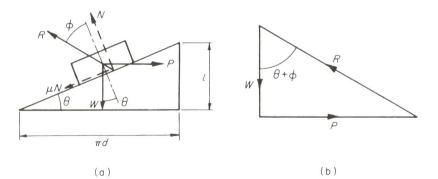

(a)   (b)

Figure 8.3   Forces on square-threaded screw

The efficiency of the screw-and-nut pair can conveniently be determined by considering the work done in one complete turn, assuming the load remains constant.

$$\text{Input} = P \times \pi d$$

$$\text{Output} = W \times \pi d \tan \theta$$

Hence efficiency

$$\eta = \frac{W \tan \theta}{P} = \frac{\tan \theta}{\tan(\theta + \phi)} \tag{8.4}$$

(from equation 8.2).

This relation can also be expressed in terms of the thread dimensions and the coefficient of friction, by expanding equation 8.4 and substituting from equation 8.1. The reader should verify that

$$\eta = \frac{1 - (\mu l / \pi d)}{1 + (\mu \pi d / l)} \tag{8.5}$$

8.2.2   V-threaded Screw

In this case, because of the semi-apex angle α (figure 8.4a), the normal reaction N at the thread is not in the same axial plane as P and W. The resolved part of N in the plane of P and W is $N \cos \alpha$. But the friction force is still μN, giving a force system as in figure 8.4b.

By comparison with figure 8.3 it can be seen that the expressions for the square thread can be used in the V-thread case by the substitution of an equivalent friction angle ϕ_e such that

$$\phi_e = \tan^{-1} \mu / \cos \alpha \tag{8.6}$$

that is

$$\mu_e = \mu / \cos \alpha \tag{8.7}$$

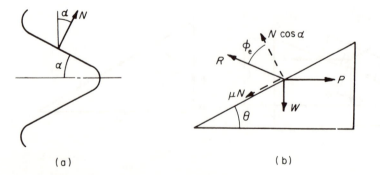

(a)   (b)

Figure 8.4   Forces on V-threaded screw

Example 8.1

The table of a planing machine has a mass of 200 kg and is driven at 0.2 m/s by a single-start square-threaded screw of mean diameter 50 mm and pitch 12 mm. Taking the coefficient of friction between the table and bed, and of the screw, as 0.15, determine the power to be applied to the screw to drive the table against a cutting tool force of 250 N.

For the screw

$$\theta = \tan^{-1}(12/50\pi) = 4°\ 22'$$

$$\phi = \tan^{-1}(0.15) = 8°\ 32'$$

Total axial load = friction resistance of table + tool force

$$= 0.15 \times 200 \times 9.81 + 250$$

$$= 544\ \text{N}$$

From equation 8.3

$$T = \tfrac{1}{2} \times 544 \times 50 \times \tan 12°\ 54'$$

$$= 3110\ \text{Nmm} = 3.11\ \text{Nm}$$

One rotation of the screw will move the table one pitch (12 mm), hence the angular velocity of the screw is

$$\omega = 2\pi \times 0.2 \times 10^3/12$$

$$= 104.7\ \text{rad/s}$$

Power $= T\omega$

$$= 3.11 \times 104.7\ \text{W} = 0.326\ \text{kW}$$

The solution can be obtained more directly by calculating the screw efficiency from equation 8.4

$$\eta = \tan 4°\ 22/\tan 12°\ 54$$

$$= 0.335$$

Power = force × velocity/efficiency

$$= 544 \times 0.2/0.335$$

$$= 326\ \text{W}$$

8.2.3   Turnbuckle

Figure 8.5 shows a simple form of turnbuckle, in which screwed members with right- and left-handed threads are drawn together against a load W by applying a torque T to the connecting member.

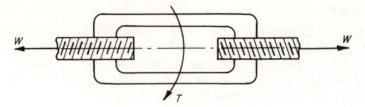

Figure 8.5   Turnbuckle

The load is clearly equal on both sides, and it follows that the torque is additive. When the lead angle of each screw is the same, then the torque is obtained by doubling equation 8.3, that is

$$T = Wd \tan(\theta + \phi)$$

It should be noted that the axial movement is also doubled, so that the efficiency remains the same as for a single screw.

8.2.4   Unloading a Screw and Nut

The friction is reversed, and the forces can be reduced to the system shown in figure 8.6 where P' is the effort required to release the load W. Then

$$P' = W \tan(\phi - \theta) \tag{8.8}$$

In fact, if $\phi < \theta$, P' will be negative, indicating that the load will run back as soon as the positive (tightening) effort is removed. If this is to be avoided it is necessary that $\phi > \theta$. The screw pair is then said to be *irreversible*.

If we take the marginal case where $\phi = \theta$, then, during tightening, equation 8.4 becomes

$$\eta = \tfrac{1}{2}(1 - \mu^2) < 0.5 \text{ whatever the value of } \mu$$

Thus screw thread irreversibility implies a theoretical maximum efficiency of

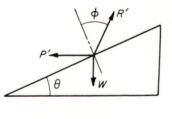

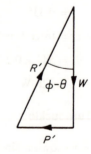

(a)

(b)

Figure 8.6   Unloading a screw

50 per cent. In practice, where irreversibility is essential, efficiencies are much less than this, often down to single figures. The following example illustrates this.

Example 8.2

Figure 8.7 shows a differential screw jack consisting of two single-start square-form right-hand screws of 48 mm outside diameter. The upper screw has a 12 mm pitch and carries the load, the lower screw has a 15 mm pitch and is attached to the base. Neither screw can rotate, and the lifting action is obtained by turning the sleeve by means of the handle.

If the coefficient of friction at the thread is 0.15 determine the least force to be applied to the handle to lift a load of 8 kN, and the overall efficiency.

Both screws are in compression under the action of the 8 kN load. Due to the differential in the pitches, the lower screw is being 'loaded' while the upper screw is being 'unloaded' (that is, the relative motion between sleeve and screw is against the load in the former case and with the load in the latter case).

Lower Screw

$$\theta = \tan^{-1}(15/48\pi) = 5°\ 40'$$

$$\phi = \tan^{-1} 0.15 = 8°\ 32'$$

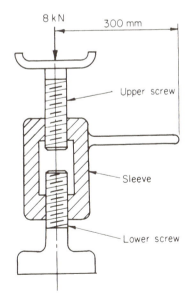

Figure 8.7   Differential screw jack

From equation 8.3

$$T = \tfrac{1}{2} \times 8000 \times 48 \tan 14° \ 12'$$

$$= 48\,600 \text{ Nmm}$$

Upper Screw

$$\theta = \tan^{-1}(12/48\pi) = 4° \ 33'$$

$$\phi = 8° \ 32'$$

From equation 8.8

$$T = \tfrac{1}{2} \times 8000 \times 48 \tan(\phi - \theta)$$

$$= 13\,370 \text{ Nmm}$$

Net torque $= 48\,600 + 13\,370 = 62\,000$ Nmm

Least force $= 62\,000/300 = 207$ N

For one complete turn of the sleeve, the load is raised by $15 - 12$, that is, 3 mm

$$\eta = \frac{\text{load} \times \text{movement}}{\text{torque} \times \text{rotation}}$$

$$= \frac{82\,000 \times 3}{62\,000 \times 2\pi}$$

$$= 0.0616 = 6.16 \text{ per cent}$$

8.3 FRICTION CLUTCHES AND BOUNDARY-LUBRICATED BEARINGS

There are many instances in machines where two surfaces, usually flat or conical in shape, are in contact under the action of a transmitted load. The two applications to be considered here are the *thrust bearing* (where one member is rotating and the other stationary), and the *friction clutch* (where both members rotate, normally in the same direction, but not necessarily at the same speed).

In a bearing the aim is to support the load and to keep the friction torque to a minimum by effective lubrication. In a clutch the requirement is to enable a high torque to be transmitted between two co-axial shafts, at the same time permitting relative slip to occur under certain conditions (for example when a speed change is made at one of the shafts in a vehicle gearbox, or for overload protection against excessive torque). As in all friction problems, the maximum torque will only be brought into play if the clutch is 'slipping' or on the point of slipping (the latter would be expected to give a somewhat higher value, but the distinction between static and kinetic friction

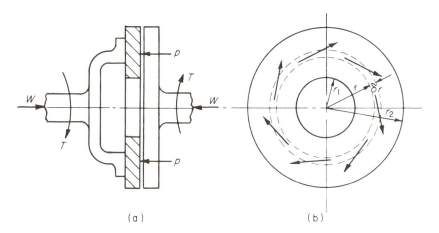

Figure 8.8   Forces and torques for flat rubbing surfaces

is not usually made in practice). When the two sides of the clutch are rotating together it behaves as a solid coupling, and will transmit any torque up to the maximum value.

In both the clutch and bearing cases the axial load W produces a pressure p at the contacting surfaces, as shown diagrammatically in figure 8.8a. Relative rotation of these surfaces sets up tangential friction forces which together have a moment about the axis of rotation equal to the transmitted torque T (as in figure 8.8b).

The relation between T and W (in terms of the dimensions and the coefficient of friction) will depend on the distribution of pressure across the sliding surfaces. For perfectly matching surfaces *uniform pressure* conditions (that is, independent of radius) would be expected to apply initially. However, when slipping takes place wear will occur, and over a long period of running this wear must be uniform for the surfaces to remain in contact. Since wear may be assumed to be proportional to the product of pressure and rubbing speed (which varies directly as the radius r), it follows that the condition for *uniform wear* is that pr is constant.

8.3.1   Flat-plate Clutches and Thrust Bearings

Due to the radial symmetry of the rubbing surfaces it is convenient to consider the forces on a thin angular ring of radius r and width δr, as in figure 8.8b.

Axial force on ring $= p2\pi r\delta r$

Total axial force on surface

$$W = 2\pi \int_{r_1}^{r_2} pr\, \mathrm{d}r \qquad (8.9)$$

Moment of tangential forces on ring $= \mu(p2\pi r\delta r)r$

Total torque on shaft

$$T = 2\pi\mu \int_{r_1}^{r_2} pr^2 \, dr \tag{8.10}$$

The integrals in equations 8.9 and 8.10 can be evaluated if the relation between p and r is known (or assumed). The two particular cases discussed in section 8.3 will now be considered.

(a) *Uniform pressure*

$$W = \pi p(r_2^2 - r_1^2)$$

$$T = 2\pi\mu p(r_2^3 - r_1^3)/3$$

Eliminating p gives the relation between torque and load

$$T = \frac{2\mu W}{3} \times \frac{r_2^3 - r_1^3}{r_2^2 - r_1^2} \tag{8.11}$$

(b) *Uniform wear*

Let $pr = k$

Then from equation 8.9

$$W = 2\pi k(r_2 - r_1)$$

and from equation 8.10

$$T = 2\pi\mu k(r_2^2 - r_1^2)/2$$

giving

$$T = \mu W(r_1 + r_2)/2 \tag{8.12}$$

Algebraically this is the simpler of the two expressions for torque, the load being effectively concentrated at the arithmetic mean radius. In consequence this form is frequently used. However, for a given axial load and dimensions, equation 8.11 gives a slightly higher torque value than equation 8.12, so it may be argued that, to be on the 'safe' side, 8.11 should be used when calculating bearing losses and 8.12 when calculating clutch torques. Unfortunately, the uncertainty involved in the value of coefficient of friction makes this argument largely academic.

The construction of a flat-plate clutch is such that two pairs of rubbing surfaces are obtained for each plate used. Figure 8.9 shows a popular type of single-plate clutch, in which the axial load is applied by means of springs and the clutch plate is free to slide on the splined end of shaft B. There is a friction 'lining' on each side of the plate and the total axial load is transmitted

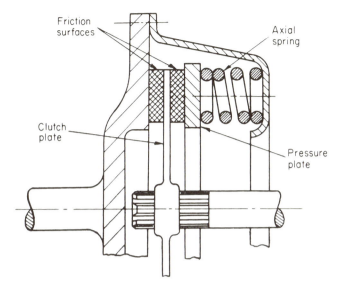

Figure 8.9   Construction of flat-plate clutch

through both pairs of surfaces. Hence the *torque* given by equation 8.11 or
8.12 *must be doubled.*

  In many applications the axial coil springs are giving way to 'diaphragm'
type springs, which are cheaper to produce and whose non-linear characteristic
results in a lower mean force for disengagement.

  Examination of equations 8.11 and 8.12 shows that, for a given outer
clutch lining radius r_2, the torque that can be transmitted increases with
increase in the inner radius r_1. Of course, this will reduce the bearing area
and hence increase the pressure for a given axial load. The pressure that can
be tolerated is limited by the amount of slip, which leads to both wear and
heat generation, but higher torque capacity can be obtained by means of
multi-plate clutches. If there are n plates, the theoretical torque is multiplied
by $2n$.

  A typical clutch for light automotive applications will have two friction
surfaces, with a ratio r_1/r_2 of between 0.6 and 0.7, and its torque capacity
will be several times greater than the maximum steady-state engine torque.
The difference between the torque capacities given by equations 8.11 and
8.12 for this application is between 1 and 2 per cent.

Example 8.3

Two co-axial shafts A and B are connected by a single-plate friction clutch
of internal diameter 130 mm and external diameter 200 mm. It may be

assumed that the normal pressure is uniform and that the coefficient of friction is equal to 0.3.

A is rotating at a constant speed of 200 rev/min and B, which carries masses having a moment of inertia of 25 kg m^2, is initially stationary with the clutch disengaged. When the clutch is engaged it takes 6 sec for B to reach full speed. Find (a) the total axial spring force and pressure, and (b) the energy dissipated during the slip period.

(a) The angular acceleration of B is constant and given by speed/time, that is

$$\alpha = \frac{200 \times 2\pi}{60 \times 6} = 3.49 \ \text{rad/s}^2$$

$$T = I\alpha$$

$$= 25 \times 3.49 = 87.2 \ \text{Nm}$$

$$= 2 \times \frac{2 \times 0.3 \times W}{3 \times 10^3} \times \frac{100^3 - 65^3}{100^2 - 65^2}$$

(from equation 8.11 doubled)

$$= 0.0502 \ W$$

Hence

$$W = 1737 \ \text{N}$$

But

$$W = \pi p (r_2^2 - r_1^2)$$

therefore

$$p = \frac{1870 \times 10^6}{\pi (100^2 - 65^2)} = 95\,700 \ \text{N/m}^2$$

(b)

Energy dissipated = work done against friction

$$= T \times \text{angle of slip}$$

$$= T \times \text{mean slipping speed} \times \text{time}$$

$$= 87.2 \times (100 \times 2\pi/60) \times 6$$

$$= 5480 \ \text{J}$$

Example 8.4

A single-plate friction clutch is to transmit 35 kW at 2000 rev/min. Assume uniform wear conditions and a coefficient of friction of 0.28. If the pressure is to be limited to 10^5 N/m^2 and the maximum diameter of the friction lining is to be 300 mm, find the minimum diameter and the axial load.

From the power

$$T = \frac{35\,000 \times 60}{2\pi \times 2000} = 167 \text{ Nm}$$

For uniform wear

$$pr = k = 10^5 r_1$$

(since maximum pressure occurs at minimum radius) and

$$T = 2\pi \mu k (r_2^2 - r_1^2)$$

(allowing for two sides) hence

$$167 = 2\pi \times 0.28 \times 10^5 r_1 \,(0.15^2 - r_1^2)$$

or

$$1000 r_1 \,(0.0225 - r_1^2) = 0.950$$

By trial

$$r_1 = 0.047 \text{ m}$$

that is, minimum diameter $= 94$ mm. Axial load

$$W = 2\pi k (r_2 - r_1)$$

$$= 2\pi \times 10^5 \times 0.047 \times 0.103$$

$$= 3030 \text{ N}$$

8.3.2 Conical Clutches and Bearings

Referring to figure 8.10 it can be seen that the annular area over which the normal pressure acts is now $2\pi r \delta r / \sin \alpha$, where α is the semi-apex angle of

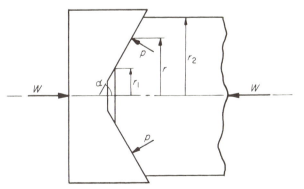

Figure 8.10 Conical clutch or bearing

the cone. Integrating the product of pressure times area and resolving axially gives

$$W = \int_{r_1}^{r_2} p2\pi r(dr/\sin \alpha) \sin \alpha$$

$$= 2\pi \int_{r_1}^{r_2} pr \, dr$$

The friction forces are obtained by multiplying the normal force on the element by μ. Integrating the moment of these forces about the axis gives

$$T = \int_{r_1}^{r_2} \mu p2\pi r(dr/\sin \alpha)r$$

$$= \frac{2\pi\mu}{\sin \alpha} \int_{r_1}^{r_2} pr^2 \, dr$$

By comparison with equation 8.10 it can be seen that the flat-plate formulae can be applied to the conical case if an equivalent coefficient of friction of $\mu/\sin \alpha$ is used.

8.4 BELT AND ROPE DRIVES

Power can be transmitted from one shaft to another by means of belt or rope drives running between pulleys attached to the shafts. Usually the shafts will be parallel, but non-parallel drives can be accommodated by means of intermediate guide pulleys.

The operation of belt and rope drives depends entirely on the development of friction forces round the arc of contact with the pulleys. The elasticity of the belt is also an important characteristic of the drive. As a method of transmitting rotational motion and power the following features should be considered.

(a) Because of the frictional nature of the drive, it is non-positive, some slip or creep between belt and pulley always occurring under load.
(b) The drive is protected against overload by the maximum tension that can be sustained by the friction forces. This protection also serves to prolong the life of the belt.
(c) Except in very light applications, such as record turntable drives, the speed ratio between the shafts is limited to about 3:1 because of the need to maintain an adequate arc of contact round the smaller pulley (though it may be possible to increase the contact by introducing an intermediate pulley).
(d) The power transmitted by any particular drive is limited by the inertia

effect on the curved lengths of belt at high speeds. For this reason values of belt speed rarely exceed 30 m/s in practice.

(e) The length of belt can easily be designed to suit a wide range of shaft centre distances.

(f) The set-up and alignment of the pulleys is not usually critical (though gross misalignment will lead to rapid belt wear), but it may be necessary to provide some adjustment for initial tensioning.

(g) Repeated flexing of the belt running on to and off the pulleys results in fatigue failure of the material, with a life usually less than 1000 hours in the case of fabric belts. However, these belts are relatively cheap and easy to replace.

The rims of the drive pulleys may be flat (or in practice slightly 'crowned' to give a self-centring action to the belt), or V-grooved. Belts may be flat (that is, rectangular section), or of wedge section (V-belt) or circular (these are referred to as ropes). The performance of the drive depends more on the pulley type than the belt section; any belt or rope driving on a flat pulley will be subject to the same force conditions, and any V-belt or rope driving in a V-groove will behave similarly (provided always that contact is with the sides of the groove and not the bottom). The analysis of these two types will now be treated separately.

8.4.1   Flat Pulley Drives

Figure 8.11a shows a typical belt drive in which the smaller pulley of radius r is the driver. The higher tension is T_1 (on the 'tight' side), and the lower tension is T_2 (on the 'slack' side), θ being the arc of contact (also called 'angle of lap').

In order to derive an expression for the maximum tension ratio it is necessary to assume that limiting friction conditions exist at all points around the arc of contact. Figure 8.11b shows the forces acting on a small length of belt subtending an angle $\delta\theta$ at the centre. As the belt passes round the pulley it is subjected to an acceleration v^2/r towards the centre (where v is the linear

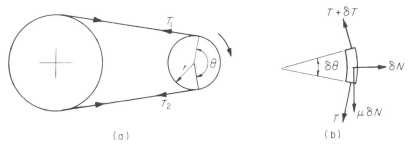

(a)   (b)

Figure 8.11   Belt drive

belt speed, equal very nearly to the pulley peripheral speed). Let m be the mass per unit length of belt.

Resolving radially

$$(T + \delta T)\sin\tfrac{1}{2}\delta\theta + T\sin\tfrac{1}{2}\delta\theta - \delta N = (mr\delta\theta)v^2/r$$

or, approximating

$$(T - mv^2)\delta\theta = \delta N \tag{8.13}$$

Resolving tangentially

$$(T + \delta T)\cos\tfrac{1}{2}\delta\theta - T\cos\tfrac{1}{2}\delta\theta - \mu\delta N = 0$$

or

$$\delta T = \mu\delta N \tag{8.14}$$

Eliminating δN between equations 8.13 and 8.14 and separating variables

$$\delta T/(T - mv^2) = \mu\delta\theta$$

Integrating between the limits of T_1 and T_2

$$\log_e \frac{T_1 - mv^2}{T_2 - mv^2} = \mu\theta$$

or

$$\frac{T_1 - mv^2}{T_2 - mv^2} = e^{\mu\theta} \tag{8.15}$$

In this expression mv^2 is sometimes referred to as the 'centrifugal tension', and it has the effect of reducing the tension available for transmitting power.

Under static or low-speed conditions mv^2 may be omitted, so that equation 8.15 reduces to

$$T_1/T_2 = e^{\mu\theta} \tag{8.16}$$

8.4.2 V-Pulley Drives

Figure 8.12a shows a V-belt driving in the groove of a pulley with a total V angle of 2α. The normal reaction on each face of an element is $\delta N'$, and is inclined at $90° - \alpha$ to the plane of symmetry of the pulley. The component of these reactions in the plane of symmetry is therefore $2\delta N'\sin\alpha$, and the friction force is $\mu2\delta N'$.

If $2\delta N'\sin\alpha$ is written δN, so that the friction force becomes $\mu\delta N/\sin\alpha$, then a direct comparison can be made between figure 8.12b and figure 8.11b. It is then seen that the relations derived for flat pulleys can be used for V-pulleys by substitution of an effective coefficient of frictional equal to $\mu/\sin\alpha$.

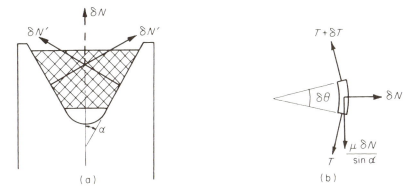

Figure 8.12   V-pulley drive

8.4.3   Conditions for Maximum Power Transmission

$$\text{Power transmitted} = (T_1 - T_2)v \qquad (8.17)$$

and, by substitution from equation 8.15

$$\text{power transmitted} = (T_1 - mv^2)(1 - e^{-\mu\theta})v \qquad (8.18)$$

Correct running tensions can only be achieved by careful setting of initial (static) tension T_0. In practice this is extremely difficult, depending as it does on length and elasticity of belt, geometry of pulleys, and centre distance. Furthermore it is rarely possible to measure this tension. However, subject to assumed values for permissible running tensions, the equations can be used to estimate power capabilities over a range of speeds.

To obtain a relation between T_1 and T_0 it may be assumed that the belt has linear elasticity and constant overall length. Consequently the stretch on the tight side (proportional to $T_1 - T_0$) must equal the contraction on the slack side (proportional to $T_0 - T_2$). Equating these gives

$$T_1 + T_2 = 2T_0 \qquad (8.19)$$

This may be written

$$T_1 - mv^2 + T_2 - mv^2 = 2(T_0 - mv^2)$$

and, by substitution for $T_2 - mv^2$ from equation 8.15 and re-arranging

$$T_1 - mv^2 = 2(T_0 - mv^2)/(1 + e^{-\mu\theta}) \qquad (8.20)$$

Power can now be expressed in terms of initial tension and speed, etc., that is, from equations 8.18 and 8.20

$$\text{power} = 2(T_0 - mv^2) \times \frac{1 - e^{-\mu\theta}}{1 + e^{-\mu\theta}} \times v \qquad (8.21)$$

For a given setting of T_0 it can be shown, by differentiation of equation 8.21 with respect to v, that the condition for maximum power is

$$T_0 - 3mv^2 = 0$$

or

$$v = \sqrt{(T_0/3m)} \tag{8.22}$$

It is sometimes argued that equation 8.18 should be differentiated to derive the condition for maximum power, but it is incorrect to assume that T_1 is a constant for a given drive. Simple substitution in equation 8.15 will show that, subject to equation 8.19, T_1 has its greatest value when $v = 0$. As v increases T_1 decreases (and T_2 increases), the power rising from zero to a maximum and then falling to zero again, as in figure 8.13.

Example 8.5

A V-belt having a mass of 0.2 kg/m and a maximum permissible tension of 400 N is used to transmit power between two pulleys of diameters 100 mm and 200 mm at a centre distance of 1 m. If $\mu = 0.2$ and the groove angle is 38°, show graphically the variation of maximum tension and power with belt speed. Hence, neglecting losses, determine the maximum power and the corresponding pulley speeds.

The limiting tension ratio is set by the angle of lap on the smaller pulley

$$\theta = \pi - 2 \sin^{-1}(50/1000) = 3.04 \text{ rad}$$

$$\exp(\mu\theta/\sin \alpha) = \exp(0.2 \times 3.04/\sin 19°) = 6.5$$

From equation 8.15

$$\frac{T_1 - 0.2v^2}{T_2 - 0.2v^2} = 6.5$$

and from equation 8.19

$$T_1 + T_2 = 2T_0$$

The maximum value of tension will be T_1 at $v = 0$. Substituting $T_1 = 400$ in equation 8.16

$$400/T_2 = 6.5$$

$$T_2 = 61.5 \text{ N}$$

and

$$T_0 = 461.5/2 = 231 \text{ N}$$

From equation 8.22 maximum power occurs at

$$v = \sqrt{(231/3 \times 0.2)} = 19.6 \text{ m/s}$$

$$\text{maximum power} = 2(231 - 77) \times \left(\frac{1 - 1/6.5}{1 + 1/6.5} \right) \times 19.6 \text{ W}$$

$$= 4.42 \text{ kW}$$

$$\text{Speed of smaller pulley} = 19.6/0.05 \text{ rad/s}$$

$$= 3750 \text{ rev/min}$$

$$\text{Speed of larger pulley} = 1875 \text{ rev/min}$$

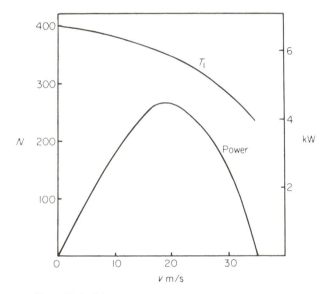

Figure 8.13   Effect of belt drive

For an intermediate value, say $v = 10$ m/s

$$mv^2 = 20$$

$$\frac{T_1 - 20}{T_2 - 20} = 6.5$$

$$T_1 + T_2 = 461.5$$

giving

$$T_1 = 385 \text{ N}$$

and, from equation 8.18

$$\text{power} = (385 - 20)(1 - 1/6.5)10 \text{ W} = 3.09 \text{ kW}$$

Zero power occurs at

$$v = \sqrt{(T_0/m)} = 34.0 \text{ m/s}$$

8.4.3   Efficiency of Belt Drives

The following are the main sources of loss arising in belt drives.

(a) *Creep.* Due to the elasticity of the belt and the change of tension round each pulley, the belt must contract as it passes over the driving pulley and extend as it passes over the driven pulley. This effect causes a 'creep' backwards round the driving and forwards round the driven pulley.

Where creep occurs, limiting friction conditions must apply, so there will be an arc (generally less than the angle of lap) over which the tension ratio is given by equation 8.15, and the remainder of the angle of lap over which the tension is constant.

It is found that the speed of each pulley corresponds to that of the belt coming on to it, that is, the driving pulley has a peripheral speed equal to the belt speed on the tight side, and the driven pulley a peripheral speed equal to the belt speed on the slack side. The percentage loss of speed between input and output is given by the difference of strain in the two sides of the belt, that is

$$\frac{(T_1 - T_2)100}{\text{belt section} \times \text{elastic modulus}} \text{ per cent}$$

In a particular case this may amount to approximately 1 per cent.

(b) *Hysteresis.* Repeated flexing of the belt between the straight and curved sections of the drive, coupled with fluctuations in tension, results in loss of energy due to internal friction in the material. This is referred to as 'hysteresis' loss, and may be of the order of 2 per cent.

8.5   BRAKES

There are three main types of mechanical brake that rely on friction over an area of contact between a stationary member and a rotating member to provide the braking torque. They are band brakes, disc brakes, and drum brakes.

8.5.1   Band Brakes

In these a flexible band (usually either a belt or a steel strip backed with friction material) is wrapped round a drum, the ends being anchored to positions that can be adjusted so as to bring the brake into operation (figure 8.14 shows a typical arrangement). Thus it is possible to allow the drum to run free or to apply a braking torque proportional to the applied force. The principle of operation is similar to that of a belt drive on a flat pulley, though now the 'belt' is stationary and the 'pulley' can rotate. The tension ratio developed round the drum is $T_1/T_2 = e^{\mu\theta}$ (as in equation 8.16).

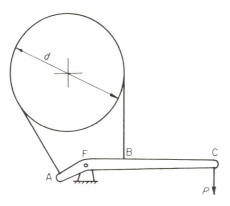

Figure 8.14   Band brake

Example 8.6

In the winding-drum band brake shown in figure 8.14 the braking force P is applied to one end of a lever ABC pivoted at F, where AF $= 25$ mm, BF $= 100$ mm and CF $= 625$ mm.

If the drum diameter is 1 m, the arc of contact 225° and the coefficient of friction 0.3, calculate the value of P required to develop a braking torque of 5000 Nm when the drum is rotating (a) clockwise, (b) anticlockwise.

The tension ratio between the ends of the belt is

$$T_1/T_2 = \exp(0.3 \times 1.25\pi) = 3.25$$

Brake torque $= (T_1 - T_2)d/2$

that is

$$5000 = (T_1 - T_2)0.5$$

$$T_1 - T_2 = 10\,000$$

Solving for T_1 and T_2 gives

$T_1 = 14\,440$ N

$T_2 = 4440$ N

(a) If the drum rotates clockwise the tight side of the belt will be at A. By moments about F for the lever

$$P \times 625 - T_2 \times 100 + T_1 \times 25 = 0$$

from above

$$25P - 4440 \times 4 + 14\,440 = 0$$

$$P = 3320/25 = 133 \text{ N}$$

(b) If the drum rotates anticlockwise the tight side will be at B. By moments about F for the lever

$$P \times 625 - T_1 \times 100 + T_2 \times 25 = 0$$

$$25P - 14\,440 \times 4 + 4440 = 0$$

$$P = 53\,320/25 = 2130 \text{ N}$$

8.5.2 Disc Brakes

The disc brake (shown diagrammatically in figure 8.15) behaves in a similar manner to a flat-plate clutch, but with a much smaller area of friction material and consequently a higher local intensity of pressure. In operation the brake pads are pressed against the disc with an equal force (applied hydraulically or mechanically) producing a torque equal to twice the product of force and effective radius (that is, the distance between the braking force and the centre of the disc).

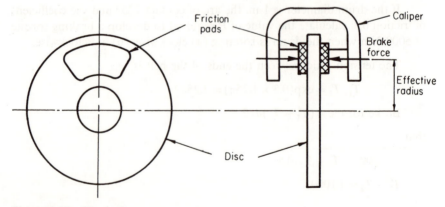

Figure 8.15   Disc brake

Example 8.7

A vehicle foot-brake has a pedal leverage of 4:1 and operates a hydraulic master cylinder of 20 mm diameter. This is connected to a disc brake caliper with pistons of 40 mm diameter at an effective radius of 80 mm. Each brake pad has an area of 1000 mm^2.

If the coefficient of friction at the pads is 0.35 and the maximum pedal effort is 400 N find the corresponding brake torque and the average pressure on the pads.

Force at master cylinder $= 400 \times 4 = 1600$ N

$$\text{Hydraulic pressure} = \frac{1600}{\pi \times 10^2} = 5.09 \text{ N/mm}^2$$

Force at each brake pad $= 5.09 \times \pi \times 20^2$

$$= 6400 \text{ N}$$

$$\text{Brake torque} = 2 \times 6400 \times 80 \text{ Nmm}$$

$$= 1024 \text{ Nm}$$

$$\text{Average pad pressure} = 6400/1000$$

$$= 6.4 \text{ N/mm}^2$$

8.5.3 Drum Brakes

Braking action is achieved by forcing a curved 'shoe', which is faced with a brake-lining material, into contact with a cylindrical drum. The shoe may rub against the inside or outside of the drum, these alternatives being called 'internal expanding' and 'external contracting' brakes. In order to minimise contact pressure and rate of wear, and to balance the applied forces, a pair of shoes is normally used.

Figure 8.16 shows a common type of internal expanding pivoted shoe brake. In this the shoes can turn about an axis O at a distance h from the drum centre. An equal braking force B is applied to each shoe at a distance H from O.

If the drum is rotating in a clockwise direction when the brake is applied then the right-hand shoe becomes a *leading shoe* (for which the friction forces tend to hold the shoe in contact with the drum) and the left-hand shoe becomes a *trailing shoe* (for which the friction forces tend to lift the shoe away from the drum). Anticlockwise rotation reverses the roles of the two shoes, so that the brake torque is independent of direction of rotation, a necessary feature for parking-brake application. If the left-hand shoe were pivoted near the top of the drum and the brake force applied from O a

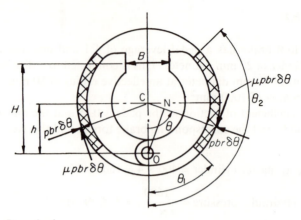

Figure 8.16   Drum brake

two-leading shoe brake would result. As will be shown this produces a higher torque for a given brake load, but only for the specified direction of rotation.

Analysis of the drum brake is more complex than for the other types discussed, depending as it does on factors such as the method of applying the load, the geometry of the shoes, and the flexibility of the lining. It is usual to assume that the drum and shoes remain rigid but the lining obeys Hooke's law and has a constant rate of wear. It can be seen from figure 8.16 that for a small rotation of a shoe about O there is a radial compression of the lining at a position θ proportional to the perpendicular ON (that is, $h \sin \theta$). For the lining to obey Hooke's law the radial pressure set up is

$$p \propto \text{ON} = kh \sin \theta$$

where k is a constant for a particular brake shoe.

The normal force on an element of lining is then $pbr \, \delta\theta$ (where b is the width of the lining) and the friction force is $\mu pbr \, \delta\theta$.

Considering the forces on each shoe (assumed to be mirror images about the central vertical) and taking moments about O

$$BH = \int_{\theta_1}^{\theta_2} pbr \, d\theta \times \text{ON} \mp \int_{\theta_1}^{\theta_2} \mu pbr \, d\theta (r - \text{CN})$$

(where the $-$ sign is for the leading shoe and the $+$ sign for the trailing shoe)

$$= kh^2 br \int \sin^2 \theta \, d\theta \mp \mu khbr \int \sin \theta (r - h \cos \theta) \, d\theta \qquad (8.23)$$

From these two cases, k_1 is determined for the leading shoe and k_2 for the trailing shoe; then the brake torque is given by

$$T = \int_{\theta_1}^{\theta_2} \mu pbr \, d\theta \times r$$

(for both shoes)

$$= \mu(k_1 + k_2)hbr^2 \int \sin \theta \, d\theta \qquad\qquad (8.24)$$

Example 8.8

In a leading/trailing pivoted shoe brake of the type shown in figure 8.16 the drum diameter is 200 mm, $h = 80$ mm, $H = 150$ mm and each brake lining extends symmetrically over 120°.

Find the value of braking force, which is the same on each shoe, to produce a torque of 1000 Nm if $\mu = 0.3$.

If the width of the lining is 40 mm what is the maximum intensity of pressure?

From equation 8.23, assuming $p = kh \sin \theta$

$$B \times 150 = k \times 80^2 \times b \times 100 \int_{30°}^{150°} \sin^2 \theta \, d\theta$$

$$\mp 0.3 \times k \times 80 \times b \times 100 \int_{30°}^{150°} \sin \theta (100 - 80 \cos \theta) \, d\theta$$

$$= 64 \times 10^4 \, kb \left[\frac{\theta}{2} - \frac{\sin 2\theta}{4} \right]_{\pi/6}^{5\pi/6}$$

$$\mp 0.24 \times 10^4 \, kb [-100 \cos \theta + 20 \cos 2\theta]_{\pi/6}^{5\pi/6}$$

$$= 0.948 \times 10^6 \, kb \mp 0.415 \times 10^6 \, kb \text{ Nmm}$$

For the leading shoe this gives

$$k_1 b = \frac{150 \, B}{0.533 \times 10^6} = 2.82 \times 10^{-4} \, B$$

For the trailing shoe

$$k_2 b = \frac{150 \, B}{1.363 \times 10^6} = 1.1 \times 10^{-4} \, B$$

From equation 8.24

$$T = 0.3(2.82 + 1.1)10^{-4} \, B \times 80 \times 100^2 \int_{30°}^{150°} \sin \theta \, d\theta$$

that is

$$1000 \times 10^3 = 0.3 \times 3.92B \times 80 \times 1.732$$

$$B = 6140 \text{ N}$$

Maximum pressure occurs on the leading shoe at $\theta = 90°$

$$p_{max} = k_1 h$$
$$= (2.82 \times 10^{-4}/40) \times 6140 \times 80$$
$$= 3.46 \text{ N/mm}^2$$

SUMMARY

The nature of boundary and film lubrication

Screw friction

Loading a square thread

Axial load $P = W\tan(\theta + \phi)$

Torque $T = Wd\tan(\theta + \phi)/2$

Efficiency $\eta = \dfrac{\tan\theta}{\tan(\theta + \phi)} = \dfrac{1 - (\mu l/\pi d)}{1 + (\mu\pi d/l)}$

Unloading a square thread

$P' = W\tan(\phi - \theta)$ $T' = Wd\tan(\phi - \theta)/2$

For a Vee thread, replace ϕ by $\tan^{-1}(\mu/\cos\alpha)$ in the above equations.

Clutches and thrust bearings

Flat friction surfaces

	Uniform pressure	Uniform wear
Axial load W	$\pi p(r_2^2 - r_1^2)$	$2\pi k(r_2 - r_1)$
Max. torque T	$\dfrac{2\mu W(r_2^3 - r_1^3)}{3(r_2^2 - r_1^2)}$	$\mu W(r_1 + r_2)/2$

For conical friction surfaces replace μ by $\mu/\sin\alpha$ in the above equations

Belt and rope drives

Flat pulleys

$$\text{Tension ratio} = \frac{T_1 - mv^2}{T_2 - mv^2} = e^{\mu\theta}$$

Initial tension $T_0 = (T_1 + T_2)/2$

$$\text{Maximum power} = 2(T_0 - mv^2)\left[\frac{1 - e^{-\mu\theta}}{1 + e^{-\mu\theta}}\right]v$$

$$\text{at } v = \sqrt{(T_0/3m)}$$

$$\text{Creep} = \frac{(T_1 - T_2) \times 100}{\text{belt section} \times E} \text{ per cent}$$

For Vee belts, replace μ by $\mu/\sin\alpha$ in the above equations.

Drum brakes

$$T = \mu(k_1 + k_2)hbr^2 \int \sin\theta \, d\theta$$

8.6   PROBLEMS

1. The clamp shown in figure 8.17 consists of two members connected by two two-start screws of mean diameter 10 mm and pitch 1 mm. The lower member is square-threaded at A and B, but the upper member is not threaded. It is desired to apply two equal and opposite forces of 500 N on the block held between the jaws, and the coefficient of friction at the screws is 0.25. Which screw should be adjusted first and what is the torque required to tighten the second screw?

   To remove the blocks from the clamp, which screw should be loosened and what is the torque required?
   [A, 0.797 Nm; B, 0.459 Nm]

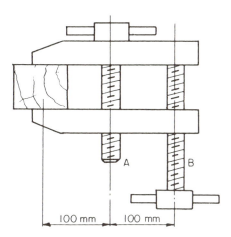

Figure 8.17

2. A machine vice consists of a fixed jaw, a movable jaw and a spindle through which there passes a tommy bar of effective length 250 mm.

   The spindle screws into the fixed jaw, the screw being square-threaded, 30 mm outside diameter and 6 mm pitch single-start. A collar on the spindle, of 40 mm mean diameter, bears on a machined surface on the movable jaw. If the coefficient of friction at the screw thread is 0.2 and at the collar is 0.1 find the minimum force on the tommy bar to produce a clamping force of 10 000 N at the jaws.
   [228 N]

3. A turnbuckle consists of a box nut connecting two rods, one screwed right-handed and the other left-handed, each having a pitch of 4 mm and a mean diameter of 23 mm. The thread is of V-form with an included angle of 55° and the coefficient of friction may be taken as 0.18. Assuming that the rods do not turn, calculate the torque required on the nut to produce a pull of 40 kN.
   [240 Nm]

4. A turnbuckle with right- and left-hand threads is used to couple two railway coaches. The threads are single-start square type with a pitch of 12 mm on a mean diameter of 40 mm. Taking the coefficient of friction as 0.1, find the work to be done in drawing together the coaches over a distance of 150 mm against a resistance which increases uniformly from 2.5 kN to 7.5 kN.
   [1550 Nm]

5. A lifting jack with differential screw threads is shown diagrammatically in figure 8.18. The portion B screws into a fixed base C and carries a right-handed square thread of pitch 10 mm and mean diameter 55 mm.

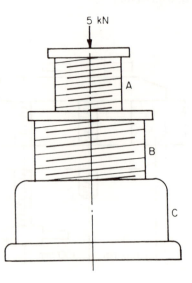

5 kN

Figure 8.18

The part A is prevented from rotating and carries a right-handed square thread of 6 mm pitch on a mean diameter of 30 mm, screwing into B. If the coefficient of friction for each thread is 0.15 find the overall efficiency and the torque to be applied to B to raise a load of 5 kN.
[9.05 per cent; 35.2 Nm]

6. A single-plate clutch has friction surfaces on each side, whose outer diameter is 300 mm. If 100 kW is to be transmitted at 1200 rev/min, and the coefficient of friction is 0.4, find the inner diameter assuming a uniform pressure of 200 kN/m².

   With these dimensions and the same total axial thrust, what would be the maximum torque transmitted under uniform wear conditions, and the corresponding maximum pressure?
   [200 mm; 784 Nm; 250 kN/m²]

7. The axial spring force operating a single-plate clutch is 8250 N. To each side of the plate is attached a ring of friction material having an inner diameter of 200 mm and an outer diameter of 350 mm. Assuming that the normal pressure p varies with radius r mm according to the relation $p(r + 50) = k$, and that $\mu = 0.3$, calculate the maximum power the clutch can transmit at 300 rev/min.
   [21.5 kW]

8. A conical clutch of included angle 60° has inner and outer radii of 25 mm and 50 mm, and the axial load is 500 N. If the variation of normal pressure with radius is given by $a + b/r$ and the maximum pressure is 1.5 times the minimum, calculate the torque transmitted for a value of $\mu = 0.35$.
   [13.3 Nm]

9. In the transmission dynamometer shown in figure 8.19, power is being transmitted from pulley A to pulley B. Pulleys C and D are mounted on a light lever pivoted at E and carrying a load W at F, where EF is 2 m. Pulleys A, C, and D are each 0.5 m diameter and B is 1.5 m diameter. A is running at 500 rev/min in an anticlockwise direction and 8 kW is being transmitted. What is the value of W?

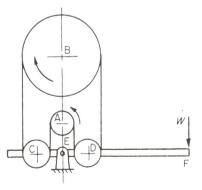

Figure 8.19

If the belt has a mass per unit length of 0.5 kg/m and the coefficient of friction between the belt and pulleys is 0.25, calculate the tension in each vertical section assuming the belt is on the point of slipping.
[306 N; 1039 N tight side, 427 N slack side]

10. An open belt connects two flat pulleys. The angle of lap on the smaller pulley is 160 degrees and the coefficient of friction is 0.25. For a given running speed, what would be the effect on the power that could be transmitted if (a) the coefficient of friction were increased by 10 per cent, or (b) the initial tension were increased by 10 per cent?
[(a) 9 per cent increase, (b) 10 per cent increase]

11. A V-pulley of 125 mm pitch diameter has grooves of 38 degrees included angle and is required to transmit 40 kW at 1200 rev/min with an angle of lap of 160 degrees. If the belt mass is 0.6 kg/m, the maximum running tension is 1000 N, and the coefficient of friction is 0.28, find the number of belts required.
[6]

12. Figure 8.20 shows a band brake in which the drum rotates anticlockwise. If the coefficient of friction is 0.4 and a force of 50 N is applied at A determine the brake torque when $r = 0.5$ m, $a = 0.8$ m, and $b = 1$ m.

What is the least value of the ratio a/r for which the brake is not self-locking?
[647 Nm; 1.557]

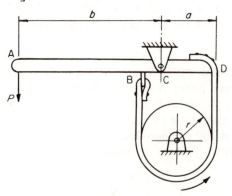

Figure 8.20

13. A four-wheel vehicle of mass 1000 kg has a foot brake with a pedal ratio of 5:1, and this actuates a hydraulic master cylinder of 25 mm diameter. The front wheels have disc brakes of effective radius 100 mm with caliper pistons of 40 mm diameter. The rear brakes are of leading/trailing shoe type, as figure 8.16, with drum diameter 200 mm, $H = 150$ mm, $h = 75$ mm. The linings subtend 90 degrees symmetrically placed and the brake cylinders are 20 mm diameter.

If the maximum pedal effort is 400 N and $\mu = 0.35$ at all brake surfaces, find the braking torque at each wheel. What is the maximum

retardation of the vehicle assuming no slip at the road wheels, which are 280 mm diameter?
[358 Nm front, 265 Nm rear; 8.93 m/s^2]

9 Toothed Gears

9.1 INTRODUCTION

Toothed gears are used to transmit motion and power between shafts rotating in a specified velocity ratio. Although there are other, and often simpler, ways of doing this (such as belt drives and friction discs) few provide the positive drive without slip and permit such high torques to be transmitted as the toothed gear. Furthermore, gears adequately lubricated operate at remarkably high efficiencies, over a very wide speed range (limited only by imperfections arising during manufacture or assembly), and between shafts whose axes are parallel, intersecting or skew.

Normally the velocity ratio is required to be constant, that is, the same in all positions of the mating gears. It is possible to design for non-uniform motion ('elliptical' gears) but their application is limited to special-purpose machinery and will not be considered here. It will be shown that the requirement for uniform motion places certain constraints on the shape of teeth that can be used for mating gears, and that further advantages following from standardisation and ease of manufacture have resulted in teeth of involute form being almost exclusively used for power gears.

The load torque that can be transmitted by gears is not limited by slip as in belt drives, but by the tooth stresses and gear materials. The distortion and deflection of the teeth under load causes secondary fluctuations in velocity ratio as they come into and out of engagement. This departure from the ideal can be partly offset by modifications to the tooth profiles, though the 'correction' will only be effective at one specified load. The result is that, in designing gears, allowance must be made for these dynamic effects by reducing the load capacity for higher running speeds.

9.2 CLASSIFICATION OF GEARS

The most appropriate grouping of gears is according to the orientation of the shafts between which they operate. These can rotate about

(1) parallel axes
(2) intersecting axes, or
(3) skew axes

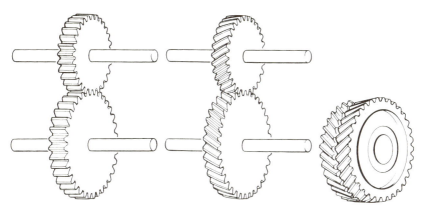

Figure 9.1   Gears for parallel axes

The first group is by far the most common, and the motion transmitted is equivalent to the rolling without slip of two circular discs. These equivalent circles are called *pitch circles*. The gears themselves have either axial teeth (when they are called *spur* gears), or helical teeth (*helical* gears). These types are shown in figure 9.1. The advantage of helical gears lies in the gradual engagement of any individual tooth, starting at the leading edge and progressing across the face of the gear as it rotates. This results in reduced dynamic effects and attendant noise, and generally prolonged life or higher load capacity for given overall size and gear material. It should be noted, however, that the single helical gear will produce an axial thrust on the shaft bearing due to the inclination of the teeth; if objectionable, this can be eliminated by employing double helical gears with two sets of teeth back to back, each set cut to opposite hand.

If the shafts are intersecting, the equivalent rolling surfaces are conical, and when teeth are formed on these surfaces they are called *bevel* gears. The teeth may be cut straight as in figure 9.2a, or curved (giving the same advantage as helical teeth). When the axes are at right angles the larger gear is called a crown wheel and the smaller a pinion.

Skew or *spiral* gears (which are helical gears of differing helix angles forming a mating pair) are used to transmit motion between non-intersecting shafts (figure 9.2b). The gears must be fixed to the shafts so that the sum of the pitch circle radii is equal to the shortest distance between the shaft axes. This is the only line joining the axes that is perpendicular to both, and this

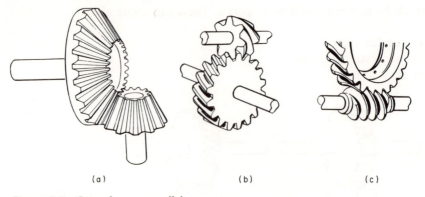

(a)   (b)   (c)

Figure 9.2   Gears for non-parallel axes

becomes the centre distance for the gears. The sum of the helix angles must be equal to the angle between the shafts. It can be shown that spiral gears have only point contact, and consequently their use is limited to light loads.

A special form of skew gears which has line contact is the worm and wheel pair (figure 9.2c). Usually, though not necessarily, the axes are at right angles. If a large number of teeth and a small helix angle are used on the wheel, a very high velocity ratio can be produced, This is accompanied by a high ratio of sliding to rolling motion, so that good lubrication is essential to reduce friction losses. Alternatively, unlubricated or boundary lubricated worm gears with small worm angles may be irreversible (that is, it will not be possible to drive the worm by turning the wheel) due to friction locking effects. Where worm gears are used in power drives they can be designed for high efficiency, but where irreversibility is a desirable feature (as in a steering gear) it can be shown that the efficiency will then be less than 50 per cent.

9.3   CONDITION FOR UNIFORM MOTION

Figure 9.3 shows a gear, centre A, rotating clockwise at an angular velocity ω_A and driving a second gear, centre B, at an angular velocity ω_B anticlockwise.

At the instant shown a pair of teeth are in contact at N. Then the point N on the gear A has a velocity

$v_{NA} = AN\omega_A$ perpendicular to AN

and the coincident point N on the gear B has a velocity

$v_{NB} = BN\omega_B$ perpendicular to BN

Now there will be a common tangent, and a common normal perpendicular to it, to the tooth profiles at N. Let LM be this common normal, where AL

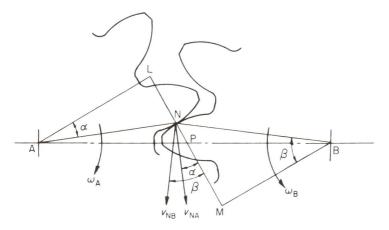

Figure 9.3   Condition for uniform motion

and BM are perpendiculars on to it. The condition for continuity of contact to be maintained as the gear rotates is that the velocity components of N along the common normal are equal, or

$$v_{NA} \cos \alpha = v_{NB} \cos \beta$$

that is

$$AN\omega_A \cos \alpha = BN\omega_B \cos \beta$$

from above. Rearranging

$$\frac{\omega_A}{\omega_B} = \frac{BM}{AL} = \frac{BP}{AP} \tag{9.1}$$

by similar triangles (where P is the intersection of the common normal LM and the line of centres AB).

For the velocity ratio to be constant, equation 9.1 implies that P must be a fixed point. Hence the condition *for constant velocity ratio* is that *the common normal to the teeth at a point of contact must pass through a fixed point on the line of centres.* This point is called the *pitch point*, the motion transmitted between A and B being equivalent to that of two cylinders, radii AP and BP, touching at P and rolling without slipping.

Teeth on two mating gears that satisfy this condition are said to be *conjugate*. It can be shown that there are an infinite number of curves which, taken in pairs, have conjugate profiles; that is, for an arbitrarily chosen shape, preferably convex, for the teeth on A, it is possible to determine geometrically the conjugate profile for teeth on B. However, in practice there are only a few standard curves that possess conjugate properties, and these will be examined in more detail later.

9.3.1   Sliding between Teeth

If the velocity components along the common tangent at the point of contact are evaluated, these are $v_{NA} \sin \alpha$ and $v_{NB} \sin \beta$. The difference between these gives the velocity of sliding

$$v_s = v_{NB} \sin \beta - v_{NA} \sin \alpha$$
$$= BN\omega_B \sin \beta - AN\omega_A \sin \alpha$$
$$= MN\omega_B - LN\omega_A$$
$$= (MP + PN)\omega_B - (LP - PN)\omega_A$$
$$= PN(\omega_A + \omega_B) \qquad (9.2)$$

since $MP/LP = BP/AP$ from equation 9.1.

This shows that the velocity of sliding is proportional to the distance of the point of contact from the pitch point, and is equal to this distance multiplied by the relative angular velocity (for an internal–external pair the numerical difference of angular velocities would be correct). Sliding velocity is an important consideration in lubrication and wear effects, and it marks a fundamental difference in character between the kinematics of toothed gears and their equivalent rolling surfaces.

9.4   CYCLOIDAL TEETH

A cycloid is the path traced out by any point on the circumference of a circle that rolls without slipping on a straight line. If it rolls on the outside of another circle it is called an *epicycloid*, and if on the inside a *hypocycloid*. These curves are shown in figure 9.4. It should be noted that I is the instantaneous centre of rotation, and hence IN is perpendicular to the cycloid at N in each case.

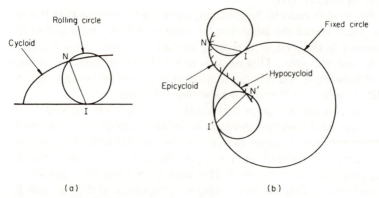

(a)   (b)

Figure 9.4   Cycloidal gears

When used for toothed gears the fixed circle becomes the pitch circle; that part of the tooth profile outside the pitch circle is an epicycloid, and that within the pitch circle is a hypocycloid. It is not essential that the rolling circles for these two portions of the teeth on a given wheel should be identical. Correct action is obtained between two mating wheels if the same size of rolling circle is used to generate the epicycloid on one wheel and the hypocycloid on the other. However, where a set of gears has to be interchangeable, any gear of the set being capable of meshing with any other, then it is necessary for a common rolling circle to be used. As long as the diameter of the rolling circle is less than the pitch circle radius, teeth will be thicker (and consequently stronger) at the root. It is therefore common practice to make this diameter equal to the radius of the smallest gear in the set (it can readily be seen that the hypocycloid for this gear follows a straight radial line).

9.5 INVOLUTE TEETH

An involute is the path traced out by a point on a line that is rolled without slipping around the circumference of a circle (called the *base circle*), as in figure 9.5a. It can also be considered as the path followed by the end of a cord being unwrapped from the base circle.

It is clear that IN, which is a tangent to the base circle, is the normal to the involute at N. When two wheels are meshing together, as in figure 9.5b, the common normal at the point of contact N becomes the common tangent LM to the base circles. Since LM intersects AB in a fixed point P, the pitch point, the condition for uniform motion is satisfied.

For the direction of rotation shown, the point of contact moves along LM during engagement. If the direction is reversed, contact takes place between the other faces of the teeth, shown dotted, and moves along L'M'. In either case the direction of normal pressure between the teeth is constant,

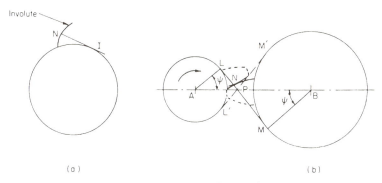

(a)   (b)

Figure 9.5   *(a) Involute curve; (b) involute teeth*

being defined by the *pressure angle* (or *angle of obliquity*) ψ. This is preferable to the variable pressure angle between cycloidal teeth, and involute gears possess a further advantage in that the setting of the centre distance is no longer critical for correct tooth action. Slight variations in the centre distance of a pair of gears will affect the amount of backlash between the teeth, and also the pressure angle, but not the uniformity of the motion transmitted.

If the base circle is enlarged to an infinite radius, its circumference becomes a straight line, and involute teeth formed from it will have straight sides. These are called *rack* teeth (figure 9.6c) and their simple shape is of particular significance in the manufacture of involute gears. A variety of 'generating' processes can be used to cut the teeth of a gear from a disc type 'blank'. The accuracy of the gear teeth depends on the accuracy of the cutter profiles (and of course on the precision of the machine), and the simple geometry of the involute rack facilitates the accurate grinding of the cutters.

9.6   INVOLUTE GEAR GEOMETRY

The desirability of establishing accepted standards for gear dimensions and tooth proportions has already been discussed. It will be necessary first to define the quantities involved and to see how these affect the motion and forces transmitted. Only the case of involute teeth will be considered in detail, though the definitions below are applicable to any tooth shape.

9.6.1   Definitions

It will be appreciated from the way in which an involute is developed that the whole of the working surface of each tooth lies outside the base circle (though some material is usually cut away at the root to provide clearance for the tips of mating teeth as they pass through the pitch point). The pitch circle divides the tooth height into an *addendum*, defined as the radial length of tooth from the pitch circle to the tip, and a *dedendum*, defined as the radial length of tooth from the pitch circle to the root. These dimensions are shown in figure 9.6.

The *circular pitch p* is the length of arc measured round the pitch circle between corresponding points on adjacent teeth. It is related to the diameter *d* and the number of teeth *t* by

$$p = \pi \frac{d_A}{t_A} = \pi \frac{d_B}{t_B} \tag{9.3}$$

and must, of course, be the same for each wheel of a pair.

For manufacturing and assembly purposes the diameter needs to be an exact number, and hence the circular pitch is not a convenient practical standard to use. The alternative is to define the tooth proportions and spacing

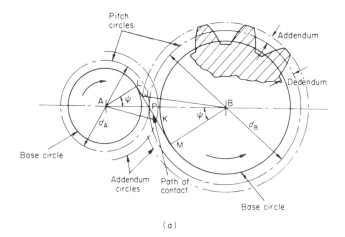

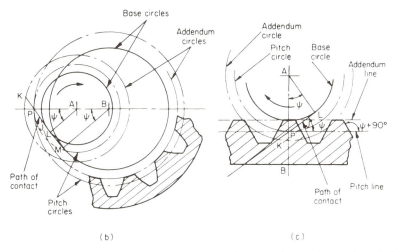

Figure 9.6   *(a) External pair; (b) external–internal pair; (c) rack and pinion*

by means of the *module m*, where

$$m = \frac{d_A}{t_A} = \frac{d_B}{t_B} \tag{9.4}$$

It is conventional to measure the module in mm.

9.6.2   Path of Contact of Spur Gears

The limits of contact between any one pair of involute teeth are determined by the intersection of the addendum circles with the common tangent to the base circles. As shown in figures 9.6a and 9.6b, LM is the common tangent, where AL and BM are inclined at the pressure angle ψ to the line of centres AB. The point M does not appear on figure 9.6c because the rack is equivalent to a wheel of infinite radius. Contact in this case lies along LP and its continuation.

First contact occurs at J where the root of a tooth on the driver meets the tip of a tooth on the driven wheel, and last contact at K when the tip of the driving tooth loses touch with the root of the driven tooth; JK is called the *path of contact*.

Since the rotation of involute gears is equivalent to a crossed belt drive between the base circles (see section 9.5 and figure 9.5b), the angle turned through by either wheel while tooth contact moves from J to K can be obtained by dividing JK by the corresponding base circle radius, that is

$$\frac{JK}{\frac{1}{2}d_A \cos \psi} \qquad \text{for wheel A}$$

and

$$\frac{JK}{\frac{1}{2}d_B \cos \psi} \qquad \text{for wheel B} \tag{9.5}$$

The distance moved by any point on the circumference of the pitch circle while contact moves from J to K is given by

$$\text{angle turned through} \times \text{corresponding pitch circle radius} = \frac{JK}{\cos \psi} \tag{9.6}$$

from equation 9.5. This is called the *arc of contact*, being the same for each wheel.

For continuity of contact it is essential that before a leading pair of teeth loses contact, the next following pair should have come into engagement. This condition requires that the arc of contact exceeds the circular pitch. In fact it would help to share the load if the arc of contact were greater than twice the circular pitch, because then there would always be at least two pairs of teeth in mesh at any time. In practice, however, it is rarely possible to achieve such a length of contact due to limitations on tooth height to avoid 'interference' (section 9.6.4). For any particular case the average number of pairs of teeth in contact, called the *contact ratio*, is given by

$$\frac{\text{arc of contact}}{\text{circular pitch}} = \frac{JK}{p \cos \psi} \tag{9.7}$$

This number will usually lie between 1.5 and 1.8, indicating that in some positions there is only one pair in contact but for more than half the time there are two pairs in contact. Since single pair contact occurs in the region nearest the pitch point, where the teeth are at their strongest, this situation is acceptable.

9.6.3 Calculation of Path of Contact

The relations between the gear dimensions and the path of contact can be determined from figure 9.6. Each case differs slightly and will be dealt with separately.

(a) *External gears*

$$JK = JP + PK$$

$$= JM - PM + LK - LP$$

$$= \sqrt{[(\tfrac{1}{2}d_B + \text{add.})^2 - (\tfrac{1}{2}d_B \cos \psi)^2]} - \tfrac{1}{2}d_B \sin \psi$$

$$+ \sqrt{[(\tfrac{1}{2}d_A + \text{add.})^2 - (\tfrac{1}{2}d_A \cos \psi)^2]} - \tfrac{1}{2}d_A \sin \psi \qquad (9.8)$$

(b) *External–internal pair*

$$JK = JP + PK$$

$$= MP - MJ + LK - LP$$

$$= \tfrac{1}{2}d_B \sin \psi - \sqrt{[(\tfrac{1}{2}d_B - \text{add.})^2 - (\tfrac{1}{2}d_B \cos \psi)^2]}$$

$$+ \sqrt{[(\tfrac{1}{2}d_A + \text{add.})^2 - (\tfrac{1}{2}d_A \cos \psi)^2]} - \tfrac{1}{2}d_A \sin \psi \qquad (9.9)$$

(c) *Rack and pinion*   The smaller wheel of a pair of gears is often called the 'pinion', and invariably so when engaging with a rack.

$$JK = JP + PK$$

$$= \text{add.}/\sin \psi + \sqrt{[(\tfrac{1}{2}d_A + \text{add.})^2 - (\tfrac{1}{2}d_A \cos \psi)^2]} - \tfrac{1}{2}d_A \sin \psi \qquad (9.10)$$

Example 9.1

A pinion of 16 teeth drives a wheel of 50 teeth at 800 rev/min. The pressure angle is $20°$ and the module is 10 mm. If the addendum is 12 mm on the pinion and 8 mm on the wheel, calculate the maximum velocity of sliding, the arc of contact, and the contact ratio.

The pitch circle diameters are obtained from equation 9.4

$$d_A = 10 \times 16 = 160 \text{ mm}$$

$$d_B = 10 \times 50 = 500 \text{ mm}$$

Then, from equation 9.8

$$JP = \sqrt{[(250 + 8)^2 - (250 \cos 20°)^2]} - 250 \sin 20° = 21.0 \text{ mm}$$

$$PK = \sqrt{[(80+12)^2 - (80\cos 20°)^2]} - 80\sin 20° = 25.7 \text{ mm}$$

$$\omega_A = \frac{50}{16} \times 800 \times \frac{2\pi}{60} = 262 \text{ rad/s}$$

$$\omega_B = 800 \times \frac{2\pi}{60} = 83.8 \text{ rad/s}$$

From equation 9.2

maximum velocity of sliding $= 25.7 \, (262 + 83.8) \text{ mm/s} = 8.88 \text{ m/s}$

From equation 9.6

$$\text{arc of contact} = (JP + PK)/\cos 20° = 46.7/\cos 20°$$

$$= 49.7 \text{ mm}$$

$$\text{circular pitch } p = \pi \times 160/16 = 31.42 \text{ mm}$$

From equation 9.7

$$\text{contact ratio} = 49.7/31.42 = 1.58$$

9.6.4   Interference

The desirability of a high value of contact ratio has been discussed in section 9.6.2, and to achieve this the path of contact should be as long as possible. However, since the whole of the involute profiles lie outside the two base circles of a mating pair, correct tooth action can only occur for contact along the common tangent external to the base circles. This condition puts a limit on the size of addendum that can be used on each wheel. For an external pair the limits are reached when the addendum circles pass through L and M in figure 9.6a. *Interference* arises at the tips of the teeth if these are exceeded. For no interference, then

$$JP \leqslant LP \qquad\qquad\qquad (9.11)$$

and

$$PK \leqslant PM \qquad\qquad\qquad (9.12)$$

It can easily be verified from the geometry that, if the addenda are equal on both wheels (as is often the case), interference is more likely to occur at the tips of the teeth on the larger wheel. This can be checked from equation 9.11. Equation 9.12 is of less significance, and clearly does not apply to the internal gear or rack cases.

It is important to ensure that interference is avoided under running conditions (or non-uniform motion will result), and also that it does not occur during manufacture (when the cutter would remove material from the involutes near the base circle of the gear, leaving 'undercut' teeth).

Experience has shown that standard tooth proportions can be laid down that will give satisfactory results over a wide range of gear sizes and numbers of teeth. Special treatment may be required for small tooth numbers or where a high gear ratio is to be obtained. Consideration will now be given to choice of standard.

9.6.5   Standard Tooth Proportions

In order to promote interchangeability of gears and economy of production it is desirable to adopt standard values for the pressure angle and tooth proportions. It has been shown that the contact ratio is determined mainly by the addendum height of the teeth, this being limited by the risk of interference occurring, particularly at the tips of teeth on the larger wheel of a pair. A high value of pressure angle will, by increasing the permissible length of contact outside the base circles, help in avoiding interference. However, because tooth profiles are made up of involutes formed from opposite sides of the base circle, they become markedly pointed at the tips when ψ approaches $30°$. Higher pressure angles are also accompanied by higher tooth and bearing loads for a given torque transmitted, due to the obliquity of tooth reactions. A suitable compromise between these factors has to be accepted.

Whereas the earliest standard, drawn up by Brown and Sharpe, used a pressure angle of $14\frac{1}{2}°$, it was realised later that this was unsatisfactory for gears with small tooth numbers. The most widely used value today is $20°$, combined with an addendulum of 1 module and a dedendum of 1.25 modules (to give the necessary clearance). The British Standard (BS 436) is based on these proportions, and the basic rack form is shown in figure 9.7.

The least number of teeth for a pinion engaging with this rack without interference is 18 (see example 9.2 below). Where it is desired to use a smaller number of teeth, interference can be eliminated by decreasing the addendum of the rack (or larger wheel of a pair of gears) and making a corresponding increase to the pinion addendum. This 'correction' to the teeth maintains the working depth at a standard 2 modules and provides an adequate contact ratio.

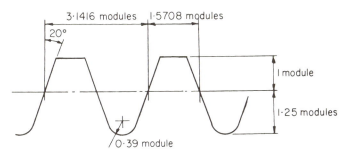

Figure 9.7   British standard basic rack

Example 9.2

Show that the maximum addendum for a rack engaging with a pinion of diameter d without interference is $\frac{1}{2}d\sin^2\psi$. If a pressure angle of 20° and an addendum of 1 module are used, determine the minimum number of teeth on the pinion.

If the pinion is to have 15 teeth and a module of 8 mm, calculate suitable addenda to give a working depth of 2 modules, and find the corresponding value of contact ratio.

For no interference at the tips of the rack teeth JP $\leqslant$ LP in figure 9.6c, that is

$$\frac{\text{add.}}{\sin\psi} \leqslant \frac{1}{2}\sin\psi$$

or

$$\text{add.} \leqslant \frac{1}{2}\sin^2\psi$$

If add. $= 1$ module $= d/t$, and $\psi = 20°$ then the inequality becomes

$$t \geqslant 2/\sin^2 20° = 17.1$$

Minimum number of teeth $= 18$

A pinion with 15 teeth and a module of 8 mm will have a pitch circle diameter of 15×8, that is, 120 mm. The maximum rack addendum that can be used is

$$\tfrac{1}{2}d\sin^2\psi = \tfrac{1}{2} \times 120 \times \sin^2 20°$$

$$= 7.0 \text{ mm}$$

To maintain a working depth of 2 modules, the pinion addendum must be 9.0 mm (note that the teeth are 'corrected' by 1 mm from the standard addendum of 1 module).

The path of contact can now be determined from equation 9.10

$$\text{JK} = 7/\sin 20° + \sqrt{[69^2 - (60\cos 20°)^2]} - 60\sin 20° = 39.8 \text{ mm}$$

The circular path is 8π mm, and hence from equation 9.7

$$\text{contact ratio} = \frac{39.8}{8\pi\cos 20°} = 1.68$$

9.7   VELOCITY AND TORQUE RATIOS

From the consideration of the condition for uniform motion, equation 9.1, and the involute gear geometry shown in figure 9.6, it is clear that the velocity ratio is given by

$$G = \frac{\omega_A}{\omega_B} = \frac{d_B}{d_A} = \frac{t_B}{t_A}$$

Strictly speaking, a pair of externally meshing gears on parallel shafts will give a negative value for G, requiring one of the diameters and corresponding tooth number to be considered as negative. In practice, the sign of the velocity ratio is often best determined by inspection. Rather more care needs to be exercised where the shafts are not parallel.

If the torques on the two gears are T_A and T_B, and losses due to sliding are ignored, then for constant angular velocity

$$\frac{T_A}{d_A/2} = \frac{T_B}{d_B/2}$$

hence

$$G = \frac{\omega_A}{\omega_B} = \frac{T_B}{T_A} = \frac{t_B}{t_A} \tag{9.13}$$

In many applications, losses due to sliding are small and equation 9.13 may be used directly with little loss of accuracy. Where losses need to be considered, the analysis in section 9.9 may be used.

A more detailed discussion of velocity and torque ratio is given in chapter 10.

9.8   GEAR TOOTH STRESSES

The models used earlier in this chapter are quite satisfactory for analysis of external effects such as torque relationships and shaft bearing loads but, for studying gear tooth strength and life, additional factors must be taken into account. These include the elastic properties of the teeth, and manufacturing and assembly tolerances. Analysis of this kind tends to be done by, or in cooperation with, specialist gear manufacturers, and much of this work depends on empirical factors derived from extensive testing. A complete treatment is beyond the scope of this book, but the concepts involved are relatively straightforward.

Gear failure can occur for a variety of reasons; the three most common are tooth fracture, fatigue of the tooth surfaces and wear. The first mode is usually catastrophic; the second and third are generally progressive, may be connected, and are frequently accompanied by secondary effects such as noise, vibration and excessive heat generation.

A number of important assumptions are made in developing a model for gear failure analysis. The more important of these are

(a) Only the tangential component of the tooth contact force is considered.
(b) The tangential component is uniformly distributed across the face width.
(c) A unity contact ratio is used, so that the greatest tooth bending stress occurs when contact is at the tip.

9.8.1 Bending Strength

For analysis of bending stress, the tooth is treated as a cantilever and the maximum stress assumed to occur at the root. Empirical factors based on gear type and size are also included. The allowable stress is based on material properties modified by a further empirical factor which is a function of pitch line velocity. In most cases both gears of a pair must be examined, but generally one or other will control the design. For the same material properties, the smaller gear will have the weaker teeth.

9.8.2 Fatigue Strength

Inaccuracies in the tooth profile, shaft spacing and alignment, and tooth deflections under load will result in small variations of angular velocity about the mean, and hence lead to dynamic loads with the attendant risk of fatigue failure. Once again, empirical factors are used, based on pitch line velocity, method of manufacture and stress concentrations, to obtain a fatigue stress which can be compared with an allowable stress.

9.8.3 Wear Strength

For mating gears, wear is primarily a function of the contact stress. Wear may take the form of *scoring* due to lubrication failure, *pitting*, a form of surface fatigue, or *abrasion* due to the presence of foreign material. Calculation of the contact stress is tedious and must be assessed in relation to the required life and operating conditions of the gear set.

9.9 EFFICIENCY OF GEARS

It has been shown that rotation of a gear is always accompanied by sliding between the teeth, the ratio of sliding:rolling motion being proportional to the distance of the point of contact from the pitch point. Although for a well-lubricated gear friction forces are relatively small, some energy loss will arise due to the sliding action. For spur gears a method of estimating this loss is given below, and it is found to be about 1 per cent, corresponding to an efficiency of 99 per cent (even when bearing losses are taken into account a value exceeding 98 per cent can often be achieved in practice). Helical gears and bevel gears produce similar, though slightly lower, values.

In skew gears, however, sliding takes place in axial as well as transverse planes, resulting generally in a much higher friction loss. Spiral gears are likely to have an efficiency in the region of 85 to 90 per cent, whereas worm gears cover a very wide range from about 35 to 95 per cent (section 9.9.2). The lower values apply to small worm angles and a high gear ratio, and the higher values to large worm angles and a low gear ratio. A special property of worm:wheel pairs is that the drive may be irreversible; in this case the efficiency will be less than 50 per cent.

9.9.1 Efficiency of Involute Spur Gears

Figure 9.8 shows the normal force F_n and the friction force μF_n at a point of contact between a tooth on wheel A and a mating tooth on wheel B.

Consider the work done against friction as the point N moves from the pitch point P to the last point of contact K (that is, the 'recess' period).

$$\text{Friction loss} = \int \mu F_n v_s \, dt$$

$$= \mu F_n (\omega_A + \omega_B) \int PN \, dt$$

(section 9.3.1)

$$= \mu F_n (\omega_A + \omega_B) \int_0^{PK/\frac{1}{2}d_A \cos \psi \omega_A} (\tfrac{1}{2} d_A \cos \psi)\omega_A t \, dt$$

$$= \mu F_n \left(\frac{\omega_A + \omega_B}{\omega_A} \right) \frac{PK^2}{d_A \cos \psi}$$

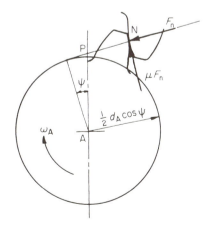

Figure 9.8   Tooth reactions

A similar expression holds during the 'approach' period for contact on the other side of P.

The corresponding useful work done during movement from P to K is F_nPK. The ratio

$$\frac{\text{friction loss}}{\text{useful work}} = \frac{\mu(\omega_A + \omega_B)}{\omega_A} \times \frac{\text{PK}}{d_A \cos \psi}$$

To obtain an approximate value consider the case of two equal wheels for which PK$/d = 0.1$, say, and the pressure angle 20°. If $\mu = 0.05$, then

$$\frac{\text{friction loss}}{\text{useful work}} = 0.05 \times 2 \times \frac{0.1}{0.94}$$

$$\approx 0.01$$

$$\text{efficiency} = \frac{\text{useful work}}{\text{total work}}$$

$$= 0.99, \text{ or 99 per cent}$$

9.9.2 Efficiency of Worm Gears

The action of a worm and wheel pair is very similar to the screw-and-nut case, and the expression for efficiency derived in section 9.2.1 may be used as a reasonable approximation, that is

$$\eta = \frac{\tan \theta}{\tan(\theta + \phi)}$$

where θ is the thread angle of the worm and ϕ the friction angle.

For a given value of ϕ it can be shown that the efficiency is a maximum when $\theta = 45° - \phi/2$. For example if $\phi = 2°$, $\eta_{max} = \tan 44°/\tan 46° = 0.933$.

On the other hand, if $\theta < \phi$ the gear is irreversible (that is, it will not be possible to drive the worm by applying a torque to the wheel). It follows that for an irreversible worm $\eta < \tan \theta/\tan 2\theta$ (and this ratio is always less than 0.5).

Example 9.3

A worm having 4 threads (that is, separate helices) and a pitch diameter of 80 mm drives a wheel having 20 teeth on a pitch diameter of 400 mm. If the effective coefficient of friction is 0.05 determine the efficiency.

The circular pitch of the wheel $= \pi \times 400/20 = 62.83$ mm

$$= \text{axial pitch of the worm}$$

The lead of the worm = pitch × number of threads

$$= 251.3 \text{ mm}$$

$$\tan \theta = \text{lead}/\text{circumference of worm}$$

$$= 251.3/80\pi = 1$$

therefore

$$\theta = 45°$$

$$\phi = \tan^{-1} 0.05 = 2° \ 52'$$

$$\eta = \frac{\tan 45°}{\tan 47° \ 52'} = \frac{1}{1.1053}$$

$$= 0.905 = 90.5 \text{ per cent}$$

SUMMARY

Condition for uniform motion

Sliding between teeth

Involute gear geometry

Circular pitch $p = \dfrac{\pi d_A}{t_A} = \dfrac{\pi d_B}{t_B}$

Module $m = \dfrac{d_A}{t_A} = \dfrac{d_B}{t_B}$

Contact ratio $= \dfrac{\text{arc of contact}}{\text{circular pitch}}$

Interference

Standard tooth proportions

Velocity and torque ratios

$$G = \frac{\omega_A}{\omega_B} = \frac{T_B}{T_A} = \frac{t_B}{t_A}$$

Efficiency

9.10 PROBLEMS

1. A pinion of 20 teeth rotating at 2000 rev/min drives a wheel of 40 teeth. The teeth are of 20° involute form with a 5 mm module and an addendum of one module.

    Determine the speed of sliding at the point of engagement, the pitch point, and the point of disengagement.

    If 10 kW is being transmitted, find the normal force between the teeth for contact at these points.

    [4.0, 0, 3.58 m/s; 510, 1020, 510 N]

2. A pinion of 30 teeth of 10 mm module meshes with a rack of 20° pressure angle. Determine the maximum addendum if it is to be the same for each, and calculate the corresponding arc of contact.

    [17.5 mm, 96.3 mm]

3. Find the length of the path of contact and the contact ratio when a pinion of 18 teeth meshes with an internal gear of 72 teeth, if the pressure angle is 20°, the module is 6 mm, and the addenda are 8 mm on the pinion and 4 mm on the wheel.

    [29.5 mm, 2.0]

4. A pair of mating gears gives a reduction of 3 to 1. The pitch diameter of the smaller gear is 80 mm, the module is 5 and the pressure angle is 20°. Find the numbers of teeth on each gear, and the contact ratio. Check for interference and suggest any necessary modifications. If the coefficient of friction between the teeth is 0.1, what is the rate of heat generation when 15 kW is supplied to the input?

    [16, 48, 1.53, interference will occur—decrease module and/or increase smaller pitch diameter, 0.59 kW]

10 Geared Systems

10.1 GEAR TRAINS

Any series of toothed gears arranged so as to transmit rotational motion from an input shaft to an output shaft is called a *gear train*.

Since most gear trains are designed to produce a speed reduction, the *gear ratio* is usually defined as the input speed:output speed ratio, thus giving a factor greater than unity. A $+$ or $-$ sign can be used to indicate the same or opposite direction of rotation.

For any pair of meshing wheels in a train, the velocity ratio is inversely proportional to the number of teeth in the wheels (and hence to the pitch circle diameters in all cases except skew and worm gears). For two external gears the direction of rotation is opposite, but for an external–internal pair it is the same.

10.1.1 Simple Trains

Figure 10.1 shows a simple train of external gears. The gear ratio is easily expressed in terms of the number of teeth

$$G = \frac{\omega_{\mathrm{A}}}{\omega_{\mathrm{C}}} = \frac{\omega_{\mathrm{A}}}{\omega_{\mathrm{B}}} \times \frac{\omega_{\mathrm{B}}}{\omega_{\mathrm{C}}}$$

$$= \left(\frac{-t_{\mathrm{B}}}{t_{\mathrm{A}}}\right)\left(\frac{-t_{\mathrm{C}}}{t_{\mathrm{B}}}\right) = \frac{t_{\mathrm{C}}}{t_{\mathrm{A}}}$$

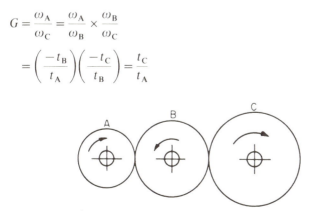

Figure 10.1   Simple gear-train

Notice that for a simple train the gear ratio is independent of the size of intermediate wheels, which serve only to take up the centre distance and to influence the direction of rotation.

In general, if there are n wheels in the train and t_o, t_i are the output and input tooth numbers

$$G = (-1)^{n-1} \frac{t_o}{t_i} \qquad (10.1)$$

The term in brackets gives the sign of the gear ratio, which could alternatively be determined by inspection—see section 9.7.

10.1.2 Compound Trains

The limitations of the simple train can be overcome, and a higher gear ratio obtained, by fixing two wheels to the intermediate shafts as in figure 10.2a. The wheels B and C are then said to be *compounded*, and rotate together with the same speed, giving

$$G = \frac{\omega_A}{\omega_D} = \frac{\omega_A}{\omega_B} \times \frac{\omega_C}{\omega_D}$$

$$= \left(\frac{-t_B}{t_A} \right) \left(\frac{-t_D}{t_C} \right) = \frac{t_B t_D}{t_A t_C} \qquad (10.2)$$

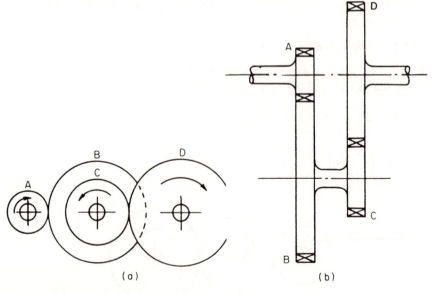

(a) (b)

Figure 10.2   Compound gear-trains

A further advantage of the compound train is that it offers the possibility of bringing the output shaft in line with the input, as in figure 10.2b. This is known as a *reverted train*, commonly found in motorcar gearboxes.

Figure 10.3 shows a schematic arrangement of a typical four-speed manual gearbox for light automotive use. The engine is connected through the main friction clutch (such as that shown in figure 8.9) to the input shaft which carries gear A. This meshes with the gear B which is integral with the layshaft. Three further gears D, F and H, also integral with the layshaft, mesh with respective gears C, E and G *which are free to rotate on the splined output shaft* when the gearbox is in neutral. When the driver selects first gear, a dog clutch (not shown) connects gear C to the output shaft. The drive then passes in sequence through the gears A, B, D and C, in exactly the same way as that shown in figure 10.2b. In a similar manner, second gear is obtained by connecting E to the output shaft, and third gear by connecting G to the output shaft. Finally, fourth or top gear is obtained by using the dog clutch to connect the input shaft directly to the output shaft, giving a straight-through drive. Conical friction clutches are usually arranged in series with the various dog clutches so that engagement of the latter does not occur until the speed difference is small. Clearly, only one dog clutch can be engaged at any one time.

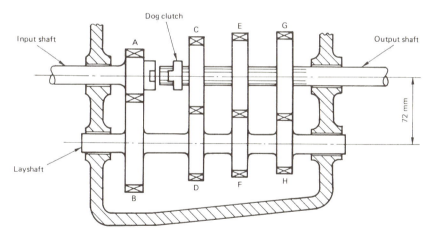

Figure 10.3   Typical motor car manual gearbox

Example 10.1

For the gear box shown in figure 10.3, find suitable tooth numbers to give drive ratios of approximately 4:1, 2.5:1, 1.5:1 and 1:1. All gears have a module of 3 and no gear is to have less than 15 teeth.

Because all wheels are to have the same pitch and a common centre-distance, each pair must have the same total of teeth, that is

$$\text{sum of diameters} = \text{twice centre-distance}$$

$$= 144 \text{ mm}$$

$$\text{sum of tooth numbers} = 144/3$$

$$= 48 \qquad (i)$$

First-gear ratio   From equation 10.2

$$G_1 = \frac{t_B t_G}{t_A t_H}$$

$$\simeq 4$$

Since from (i)

$$t_A + t_B = t_G + t_H = 48$$

it is possible to make

$$t_B = t_G = 32$$

and

$$t_A = t_H = 16$$

giving $G_1 = 4$ exactly.

Second-gear ratio

$$G_2 = \frac{t_B t_E}{t_A t_F} = 2\frac{t_E}{t_F}$$

(from above)

$$\approx 2.5$$

The nearest to this value is obtained with $t_E = 27$, $t_F = 21$, giving $G_2 = 2.58$.

Third-gear ratio

$$G_3 = 2\frac{t_C}{t_D} \approx 1.5$$

The nearest is with $t_C = 21$, $t_D = 27$, giving $G_3 = 1.55$.

Alternative solutions can be found for which $t_A = 15$ and $t_B = 33$, or for $t_A = 17$, $t_B = 31$, both satisfying the limitation on the smallest wheel size. Following up the latter alternative gives $t_G = 33$, $t_H = 15$ and $G_1 = 4.02$; $t_E = 28$, $t_F = 20$ and $G_2 = 2.57$; $t_C = 22$, $t_D = 26$ and $G_3 = 1.54$.

10.2 EPICYCLIC GEAR TRAINS

In an epicyclic gear one (or more) wheel is carried on an 'arm' which can rotate about the main axis of the train. These wheels are free to rotate relative to the arm and are called 'planets'.

In practice, to improve the dynamic characteristics, planets are used in symmetrically placed groups, generally of three or four, and the arm takes the form of a 'spider' or 'planet carrier'.

The simplest type of epicyclic consists of a 'sun' wheel S (figure 10.4) and a concentric ring gear or 'annulus' A, the planets P meshing externally with S and internally with A. Endless variations on this simple system are possible, but for the sake of classification and analysis any epicyclic having a single carrier will be called *single stage*. *Multi-stage* epicyclics can be formed by coupling two or more single-stage trains.

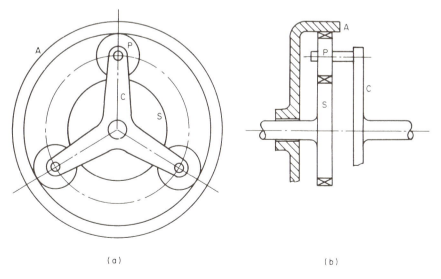

(a)   (b)

Figure 10.4 Simple epicyclic train

10.2.1   Advantages of Epicyclics

When compared with gear trains on fixed-position axes, epicyclics possess significant advantages which in certain circumstances may justify the additional complexity.

(a) They are compact in space requirements, particularly when output and input have to be in line.
(b) Both static and dynamic forces are balanced if multiple planets are used.
(c) A high torque-capacity can be achieved by the use of multiple planets.
(d) High gear-ratios can easily be achieved, although this often results in high planet speeds.
(e) A variable gear-ratio (either stepless or selective) can be achieved in a number of ways.

10.2.2   Single-stage Epicyclics

It is difficult to visualise the relative motion of the wheels in an epicyclic train because of the effect produced by the rotation of the arm. For example, in the simple epicyclic of figure 10.4 when operating with a fixed annulus, rotation of the sun will cause the planets and carrier to follow at a lower speed which must be determined from the numbers of teeth in the wheels.
  Methods of analysis depend on

(a) considering the wheel speeds relative to the arm (that is, treating the arm as initially fixed), followed by
(b) giving an equal rotation to all wheels (and the arm) about the main axis of the train as though the gears were locked solid.

  Each of these motions, although of a restricted kind, is a possible mode of rotation and consequently a combination of any proportion of (a) with (b) is also possible. By this means the required conditions can be satisfied.
  It should be noted that for any epicyclic it is necessary to specify the speed of two members (one could be zero) in order to be able to determine the speeds of all.

Example 10.2

For the epicyclic train of figure 10.4, find the gear ratio between the sun and planet carrier when the annulus is fixed, if S has 50 teeth and A 90 teeth.

Although it will not be necessary to determine the speed of the planet, this will be shown for sake of completeness. From the layout of the gear it can be seen that

$$t_S + 2t_P = t_A$$

and hence

$$t_P = 20 \qquad\qquad (i)$$

It is convenient to enter the rotations of each member in a table, which will show clearly the steps in the solution.

Operation	Rotation			
	C	S	P	A
Fix arm and give A x rev	0	$-\dfrac{90}{50}x$	$\dfrac{90}{20}x$	x
Give all y rev	y	y	y	y
Add	y	$y - 1.8x$	$y + 4.5x$	$x + y$

The condition to be satisfied (fixed annulus) is

$$x + y = 0 \qquad\qquad (ii)$$

From the table

$$G = \frac{y - 1.8x}{y}$$

$$= 2.8 \text{ (from (ii))}$$

An alternative approach is to consider the motion of the various members *relative* to the arm, that is, the motion they would have if the arm were fixed. Thus the velocity of the sun relative to the arm is $\omega_S - \omega_R$, and the velocity of the annulus relative to the arm is $\omega_A - \omega_R$. The ratio of relative velocities can then be stated using equation 10.1

$$\frac{\omega_S - \omega_R}{\omega_A - \omega_R} = -\frac{t_A}{t_S}$$

and this is always true. Putting the particular condition $\omega_A = 0$ and rearranging

$$\frac{\omega_S}{\omega_R} = \frac{t_A}{t_S} + 1 = 2.8$$

Comparison of the two methods will show that, in reality, they are the same. While the first approach is perhaps simpler to visualise, the second generally involves less work, especially for more complex systems.

One method by which the output speed can readily be varied is to drive one member of the epicyclic from an independent external wheel. Because this wheel will turn about a fixed axis it cannot be treated as part of the epicyclic and should not be included in the table of rotations. An example is given below, in which a compound planet is used, engaging with a second sun wheel in place of the annulus.

Example 10.3

In the epicyclic train shown in figure 10.5 the wheels A and E (30 teeth) are fixed to a sleeve Y which is free to rotate on spindle X. B (24 teeth) and C (22 teeth) are keyed to a shaft which is free to rotate in a bearing on arm F. D (70 teeth) is attached to the output shaft Z. All teeth have the same pitch.

The shaft X makes 300 rev/min and the shaft V 100 rev/min in the same direction. The wheel H has 15 teeth. Determine the speed and direction of rotation of Z.

$$t_A + t_B = t_C + t_D$$

therefore

$$t_A = 22 + 70 - 24 = 68$$

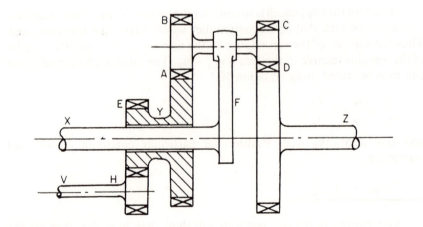

Figure 10.5

A table of speeds can now be made up (excluding H which is not part of the epicyclic). The second stage will be omitted, the general relation being obtained by adding the same quantity to each member.

	X	A, E	B, C	Z
Fix F, give A x rev/min	0	x	$-\dfrac{68}{24}x$	$\dfrac{68}{24} \times \dfrac{22}{70}x$
Add y rev/min to each	y	$x + y$	$-2.833x + y$	$0.891x + y$

The conditions to be satisfied are that X does 300 rev/min and A, E (being driven externally from V with a reduction of 2:1) do -50 rev/min.

From the table

$$y = 300$$

$$x + y = -50$$

therefore

$$x = -350$$

$$\text{speed of } Z = 0.891x + y$$

$$= -11.8 \text{ rev/min}$$

Where epicyclics contain bevel gears it is not possible to specify by a $+$ or $-$ sign the direction of rotation of the planets. Since they are only intermediate wheels it is best to omit them from the table.

The most common form of bevel epicyclic is found in the vehicle differential gear (figure 10.6). In this the planets P are carried round with the casing C which is driven from the vehicle gearbox. The action of the differential is to split the drive into two, along half-shafts 1 and 2.

	C	1	2
Fix C, give x rev to 1	0	x	$-x$
Add y rev	y	$y + x$	$y - x$

It can be seen from the table that the mean drive-speed y is equal to that of the casing, but small differences x between the two half-shafts (due to variations in road conditions, cornering, etc.) can be accommodated.

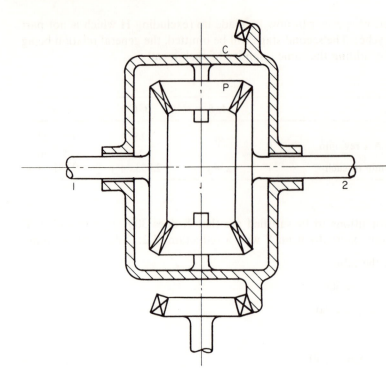

Figure 10.6   Bevel differential gear

10.2.3   Multi-stage Epicyclics

These are frequently used in automatic or pre-selective gearboxes, the planet carrier of one train being attached to (and rotating with) the sun (or annulus) of a second train. This coupling principle can be extended to any number of trains. Selection of a particular overall gear-ratio is usually achieved by fixing one member of the train by means of a clutch or brake.

Example 10.4

Figure 10.7 shows a two-stage epicyclic gear. The input shaft P is connected to sun wheels S_1 (40 teeth) and S_2 (20 teeth), and the output shaft Q is attached to the carrier for P_2. A_1 has 80 teeth and A_2, which forms the carrier for P_1, has 100 teeth.

Find the gear ratio when (a) A_1 is fixed; (b) A_2 is fixed.

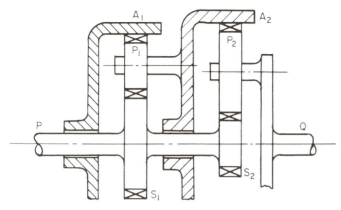

Figure 10.7   Two-stage epicyclic train

(a) Consider each train separately, starting with A_1, P_1, S_1 because it is given that A_1 is to be stationary. The table of speeds becomes

	Arm 1 (A_2)	A_1	*P (S_1, S_2)*
Fix arm, give A_1 x	0	x	$-2x$
Add $-x$ to fix A_1	$-x$	0	$-3x$

Now proceed to the train A_2, P_2, S_2.

	Q (Arm 2)	A_2	S_2
Fix arm, give A_2 y	0	y	$-5y$
Add z to each	z	$y + z$	$-5y + z$

For the speeds from the two tables to be compatible

$$y + z = -x$$

$$-5y + z = -3x$$

giving $y = x/3$ and $z = -4x/3$.

Gear ratio $P:Q = -3x:z = 2.25:1$

(b) With A_2 stationary the gear ratio is obtainable directly from the train A_2, P_2, S_2.

	Q	A_2	S_2 (P)
Fix arm, give A_2 v rev	0	v	$-5v$
Add $-v$ to fix A_2	$-v$	0	$-6v$

Giving a gear ratio of 6.

Note that A_1 will rotate freely, but does not influence the motion in this case.

Using the relative velocity approach the two equations (one for each stage) that describe the system are

$$\frac{\omega_{S1} - \omega_{R1}}{\omega_{A1} - \omega_{R1}} = -\frac{t_{A1}}{t_{S1}} \quad \text{and} \quad \frac{\omega_{S2} - \omega_{R2}}{\omega_{A2} - \omega_{R2}} = -\frac{t_{A2}}{t_{S2}}$$

For the particular configuration

$$\omega_{S1} = \omega_{S2} = \omega_P, \qquad \omega_{R1} = \omega_{A2}, \qquad \omega_{R2} = \omega_Q$$

and with the appropriate tooth numbers

$$\frac{\omega_P - \omega_{A2}}{\omega_{A1} - \omega_{A2}} = -2 \quad \text{and} \quad \frac{\omega_P - \omega_Q}{\omega_{A2} - \omega_Q} = -5$$

(a) $\omega_{A1} = 0$

Eliminating ω_{A2} and rearranging gives

$$\omega_P / \omega_Q = 18/8 = 2.25$$

(b) $\omega_{A2} = 0$

The second equation gives the solution directly

$$\omega_P / \omega_Q = 6$$

10.3   TORQUE RELATIONS IN GEARBOXES

In any gearbox transmitting power, when there is a speed change between input and output there will be a related torque change. This further implies that, for equilibrium of the gearbox, an external fixing-torque must be applied to balance the net difference between input and output torques.

Let T_i be the input torque at an angular velocity ω_i, T_o the output reaction-torque (*on* the gearbox) acting in the opposite sense to the angular velocity ω_o.

$$\text{Input power} = T_i \omega_i \tag{10.3}$$

If η is the transmission efficiency

$$\text{output power} = \eta T_i \omega_i = -T_o \omega_o$$

that is, the net power absorbed by the gearbox is zero, or

$$\eta T_i \omega_i + T_o \omega_o = 0 \tag{10.4}$$

The net torque on the gearbox is also zero, that is

$$T_i + T_o + T_f = 0 \qquad (10.5)$$

where T_f is the fixing torque on the gearbox.

Example 10.5

Find the torque on the casing of the gearbox in example 10.1 under the following conditions. Input from the engine is 20 kW at a speed of 3000 rev/min. First gear is engaged to give a reduction of 4:1 with an efficiency of 98 per cent.

From equation 10.3

$$T_i = \frac{20\,000 \times 60}{3000 \times 2\pi} = 63.8 \text{ Nm}$$

From equation 10.4

$$0.98 \times 20\,000 + T_o \times \frac{750 \times 2\pi}{60} = 0$$

that is

$$T_o = -250 \text{ Nm}$$

From equation 10.5

$$T_f = 250 - 63.8$$
$$= 186.2 \text{ Nm}$$

10.4   TORQUE TO ACCELERATE GEARED SYSTEMS

In the previous section torque relations were determined under constant-speed conditions when a known power was being transmitted. However, if the speed is not constant an additional torque will be required to accelerate the gearbox together with any associated masses or inertia on the input and output sides.

10.4.1   Acceleration of Rotational Systems

If any two shafts A and B are rotating in a fixed speed ratio $G = \omega_B/\omega_A$ it follows that the angular accelerations will also be in the same ratio, that is

$$\alpha_B/\alpha_A = G$$

Let I_A, I_B be the total moments of inertia associated with shafts A, B.

Torque on A to accelerate $I_A = I_A \alpha_A$

Torque on B to accelerate $I_B = I_B \alpha_B$

Power absorbed to accelerate the system $= (I_A \alpha_A) \omega_A + (I_B \alpha_B) \omega_B$

$$= (I_A + G^2 I_B) \alpha_A \omega_A \tag{10.6}$$

If this is produced by applying a torque T_A to A, then power $= T_A \omega_A$. Equating this to equation 10.6 gives

$$T_A = (I_A + G^2 I_B) \alpha_A \tag{10.7}$$

$I_A + G^2 I_B$ is called the *equivalent inertia* referred to shaft A. The system can then be treated as a single inertia problem, and there is no limit to the number of inertias at different speed-ratios which can be referred in this way. Care must be taken to express G as the ratio of actual speed:referred axis speed.

Example 10.6

In the epicyclic shown in figure 10.8 the wheel D (40 teeth) is attached to shaft Y which is held stationary. There are two pairs of compound planets B (30 teeth) and C (50 teeth) rotating freely on pins attached to plate Z, which can rotate independently about the axis XY. All teeth are of 6 mm module.

B and C together have a mass of 3 kg and a polar radius of gyration of 75 mm. A has a polar moment of inertia of 0.06 kg m² and Z, together with attached pins, of 0.35 kg m².

Find the torque required on shaft X to produce an acceleration of 5 rad/s² at Z.

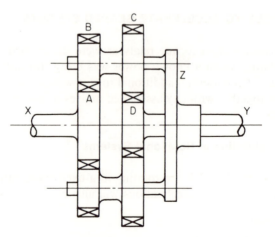

Figure 10.8

Since all teeth have the same pitch

$t_A = t_D + t_C - t_B = 60$

The speed ratios can be determined in the usual way.

	Z	D(Y)	B, C	A(X)
Fix Z, give D x rev	0	x	−0.8x	0.4x
Subtract x to fix D	−x	0	−1.8x	−0.6x
Divide by −0.6x to reduce A to unity	1.67	0	3	1

In addition to rotating about their own axis, the planets have a rotation about the main axis XY. To allow for this the moment of inertia of the plate Z must be increased by considering the mass of the planets acting at their centre of gravity (that is, the position of the pins).

Radius arm of pins on Z = radius of D + C

$$= \tfrac{1}{2}(t_D + t_C) \times \text{module}$$

$$= 270 \text{ mm}$$

Moment of inertia of plate Z and two pairs of planets

$$= 0.35 + 2 \times 3 \times 0.270^2 = 0.788 \text{ kg m}^2$$

Equivalent inertia of system referred to X

$$= 0.06 + 1.67^2 \times 0.788 + 3^2 \times 2 \times 3 \times 0.075^2$$

(from the speed ratios determined)

$$= 2.55 \text{ kg m}^2$$

Since accelerations will be in the same ratio as speeds, acceleration of X is 3 rad/s^2 when Z accelerates at 5 rad/s^2, therefore

torque on X = 2.55 × 3 = 7.65 Nm

Transfer of torque to reference axis. Consider again the two shafts A and B for which $G = (\omega_B/\omega_A) = \alpha_B/\alpha_A$, but in addition to the torque T_A applied to A let there be a resisting torque T_B (that is, in the sense opposing the direction of ω_B) acting on B.

Since the net power absorbed by the system is now $T_A\omega_A - T_B\omega_B = (T_A - GT_B)\omega_A$, equation 10.7 is modified to read

$$T_A - GT_B = (I_A + G^2 I_B)\alpha_A \tag{10.8}$$

This shows that a torque can be transferred from one shaft to another by multiplying it by the speed ratio between the shafts.

Example 10.7

A reduction gear having a speed ratio G connects an input shaft, with associated moment of inertia I_1, to an output shaft carrying moment of inertia I_o. If a constant torque T_i is applied at the input and there is a load torque T_o on the output shaft, obtain an expression for the output acceleration.

Note that $G = \omega_i / \omega_o$ and hence the net accelerating torque referred to the output shaft

$$= GT_i - T_o$$

The equivalent inertia referred to the output shaft

$$= G^2 I_i + I_o$$

The ratio of torque to inertia gives the output acceleration

$$\alpha_o = \frac{GT_i - T_o}{G^2 I_i - I_o}$$

10.4.2 Acceleration of Combined Linear and Rotational Systems

Consider a vehicle that has a total mass m, a moment of inertia of engine rotational parts I_e, and of all the wheels together I_w (some assessment will have to be made to include part of the gearbox and drive-axle inertia with either I_e or I_w).

Then if G is the gear ratio and r the radius of the road wheels, it can readily be seen that the total kinetic energy of the system when the wheel speed is ω is

$$\tfrac{1}{2}(mr^2 + G^2 I_e + I_w)\omega^2 \qquad (10.9)$$

The quantity within the brackets is the equivalent inertia of the vehicle referred to the road wheels, taking into account the linear- and rotational-energy terms (the vehicle mass has effectively been added at the wheel radius to the moment of inertia of the wheels).

Example 10.8

A vehicle of total mass 1000 kg has an effective road-wheel diameter of 720 mm. The moments of inertia of each front and rear wheel are 1.5 kg m^2 and 2 kg m^2 respectively. The moment of inertia of the engine rotational parts is 0.4 kg m^2.

The engine torque may be assumed constant and equal to 150 Nm over a wide speed range, and drives the back axle through a 5:1 reduction. Friction torque on the engine shaft is 20 Nm and total axle friction is 25 Nm. If windage and rolling resistance is $200 + v^2$ N at a speed of v m/s, calculate the acceleration of the vehicle at a speed of 15 m/s and the time taken to accelerate from 15 to 30 m/s.

Net torque referred to driving wheels

$$= G(\text{engine torque} - \text{friction torque}) - \text{axle friction} -$$

resistance × wheel radius

$$= 5(150 - 20) - 25 - (200 + v^2)0.36$$

$$= 553 - 0.36 \, v^2 \text{ Nm}$$

Equivalent inertia referred to wheels

$$= mr^2 + G^2 I_e + I_w$$

$$= 1000 \times 0.36^2 + 5^2 \times 0.4 + 7 = 147 \text{ kg m}^2$$

Angular acceleration of wheels

$$= \frac{553 - 0.36 \, v^2}{147}$$

At $v = 15$ m/s, the linear acceleration of the vehicle is

$$\frac{553 - 0.36 \times 15^2}{147} \times 0.36 = 1.15 \text{ m/s}^2$$

Multiplying angular acceleration by wheel radius gives an expression for the linear acceleration

$$= 1.35 - 0.00088 \, v^2 = \frac{dv}{dt}$$

Hence, the time to acceleration from 15 to 30 m/s

$$= \int_{15}^{30} \frac{dv}{1.35 - 0.00088 \, v^2}$$

$$= 1136 \int \frac{dv}{1534 - v^2}$$

$$= \frac{1136}{78.4} \int \left[\frac{1}{39.2 - v} + \frac{1}{39.2 + v} \right] dv$$

$$= 14.5 \left[\log_e \left(\frac{39.2 + v}{39.2 - v} \right) \right]_{15}^{30}$$

$$= 14.5 \log_e 3.36 = 17.6 \text{ s}$$

10.5 ANGULAR IMPULSE IN GEARED SYSTEMS

There are many examples in practice where two sub-systems rotating independently are then coupled together and caused to rotate in a fixed speed-ratio. The means of coupling, in order to limit the impact loading, usually incorporates a friction device such as a clutch or belt drive.

Consider first the simple system of figure 10.9a, in which I_1 is initially rotating clockwise and I_2 is stationary. Suppose that friction discs, of radius r_1 and r_2, attached to the respective shafts, are now forced together. It can be seen that the torque on I_1 is Fr_1 and on I_2 is Fr_2, and since they act for equal time on each, until slipping ceases

$$\frac{\text{change of angular momentum of } I_1}{\text{change of angular momentum of } I_2} = \frac{r_1}{r_2} \qquad (10.10)$$

Each case should be considered on its merits to determine the directions in which the speeds (and hence angular momenta) change.

It is clear that the total angular momentum of the system is *not* constant, because of the difference in angular impulse on each sub-system. Note that these torques are always in the same ratio as the *final* speeds (that is, the gear ratio), even though the speed ratio is continually changing during the slipping period. When the bearing reactions are considered, it is found that they constitute an external couple on the system equal to $F(r_1 + r_2)$, and it is this which accounts for the overall change in angular momentum.

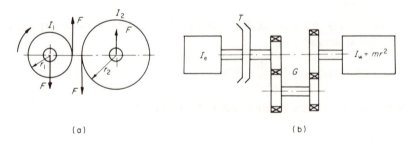

Figure 10.9   Impact of geared systems

Figure 10.9b shows the vehicle system diagrammatically, treating it as two sub-systems that can be joined through a clutch and gearbox in a fixed ratio G. If at any instant the clutch is disengaged and the engine and wheel speeds are ω_e and ω_w

angular momentum of engine $= I_e \omega_e$

'angular' momentum of wheels and vehicle $= (I_w + mr^2)\omega_w$

When the clutch is let in, a torque T (assumed constant) will be applied to the engine, and a torque GT to the wheels. The time of application of these torques is the period of clutch slip, say t, so that the angular impulses are Tt and GTt respectively, and acting in opposite senses. These can be equated to the changes in angular momenta (in this case if the engine speed falls the wheel speed will rise, and vice versa). The effect of engine torque can be neglected since the synchronising time is short.

Let ω be the wheel speed immediately after engagement, then

$$Tt = I_e(\omega_e - G\omega) \tag{10.11}$$

and

$$GTt = (I_w + mr^2)(\omega - \omega_w) \tag{10.12}$$

In such an engagement there will be a loss of energy in the system, and this can be calculated from the individual kinetic energies. This energy is dissipated in the clutch, where it is converted into heat.

Example 10.9

A car has a total mass of 750 kg. The road wheels are 670 mm diameter and each has a moment of inertia of 2 kg m^2. The moment of inertia of the engine rotating parts is 0.4 kg m^2.

With the clutch disengaged and the engine idling at 500 rev/min the car is coasting at 8.5 m/s. The bottom gear of 19:1 is then engaged and the clutch pedal released. Find the speed of the car after clutch slipping ceases, and the corresponding loss of kinetic energy.

$$\text{Initial engine speed} = 500 \times 2\pi/60 = 52.4 \text{ rad/s}$$

$$\text{Initial wheel speed} = 8.5/0.335 = 25.4 \text{ rad/s}$$

Substituting in equations 10.11 and 10.12

$$Tt = 0.4(52.4 - 19\omega)$$

$$19Tt = (4 \times 2 + 750 \times 0.335^2)(\omega - 25.4)$$

Eliminating Tt gives

$$19 \times 0.4(52.4 - 19\omega) = 92.2(\omega - 25.4)$$

$$\omega = 2738/236.5 = 11.6 \text{ rad/s}$$

$$\text{Speed of car} = 11.6 \times 0.335 = 3.88 \text{ m/s}$$

Loss of kinetic energy

$$= \tfrac{1}{2}[0.4(52.4^2 - 19^2 \times 11.6^2) + 8(25.4^2 - 11.6^2) + 750(8.5^2 - 3.88^2)]$$

$$= \tfrac{1}{2}(-18\,300 + 4080 + 42\,800)$$

$$= 14\,300 \text{ Nm}$$

SUMMARY

Simple gear trains

$$G = \frac{\omega_A}{\omega_B} = \pm \frac{t_B}{t_A} \quad \text{(sign by inspection)}$$

Compound gear trains

$$G = \frac{\omega_A}{\omega_B} \times \frac{\omega_C}{\omega_D} = \pm \frac{t_B}{t_A} \times \frac{t_D}{t_C} \quad \text{(sign by inspection)}$$

Epicyclic gear trains

Consider motion relative to the carrier, which is equivalent to simple or compound train motion plus rotation of whole assembly

$$\frac{\omega_S - \omega_R}{\omega_A - \omega_R} = \pm \frac{t_A}{t_S} \quad \text{(sign by inspection)}$$

Torque relationships

Use torque and power summations

Referred inertia

To refer an inertia to a shaft running at a difference speed, multiply by G^2 where G is the ratio actual speed to referred shaft speed

Angular impulse in geared systems

10.6 PROBLEMS

1. Two shafts A and B in the same line are geared together through an intermediate shaft C. The wheels connecting A and C have a module of 3 mm, and those connecting C and B a module of 6 mm, and no wheel is to have less than 15 teeth.

   If the ratio at each reduction is the same, and the speed of B is to be approximately but not greater than 1/12 the speed of A, find suitable wheels, the actual reduction and the centre distance to shaft C from A and B.
   [30, 104, 15, 52, 12.02, 201 mm]

2. A train of spur gears is to give a total reduction of 250:1 in four steps. No pinion is to have less than 20 teeth, and the modules are to be 5 mm, 7.5 mm, and 10 mm for the first three stages. If the system is to be as compact as possible and the input and output are co-axial, determine suitable numbers of teeth and the module for the last pair of wheels.
   [20, 100; 20, 100; 20, 80; 20, 50; 10 mm]

3. A simple sun–planet–annulus epicyclic train has 40 teeth on the sun wheel and 90 teeth on the annulus. If the speed of the sun is 900 rev/min, at what speed must the annulus be driven to cause the planet carrier to rotate at 18 rev/min (a) in the same direction as the sun, (b) in the opposite direction?
[374, 426 rev/min]

4. An epicyclic gear used as a rev counter consists of a fixed annulus A of 22 teeth, an annulus B of 23 teeth rotating about the axis of A, and an arm C also rotating about this axis. C carries wheels D and E of 19 and 20 teeth respectively meshing with A and B and fixed to each other.

   If the shaft driving C is connected to an electric motor through a reduction gear, what should this reduction be so that a recording drum attached to B rotates once per 1000 revs of the motor?
[6.86]

5. In the epicyclic gear shown in figure 10.10 the internal wheels A and F and the compound wheel C–D rotate independently about the axis O. The wheels B and E rotate on pins attached to the arm L. All wheels have the same pitch and the numbers of teeth are: B and E 18, C 28, D 26.

   If L makes 150 rev/min clockwise, find the speed of F when (a) A is fixed, (b) A makes 15 rev/min counter-clockwise.
[6.22 rev/min clockwise; 8.16 rev/min counter-clockwise]

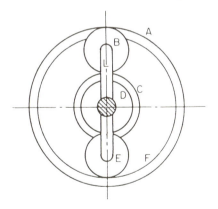

Figure 10.10

6. In figure 10.11 the driving shaft A is rotating at 300 rev/min and the casing C is stationary. E and H are keyed to the central vertical spindle and F can rotate freely on this spindle. K and L are fixed to each other and rotate on a pin fitted to the underside of F. L meshes with internal teeth on the casing.

   The numbers of teeth on the wheels are as shown. Find the number of teeth on the casing, and the speed and direction of rotation of B. H, K, L and C all have the same module.
[90; 100 rev/min opposite to A]

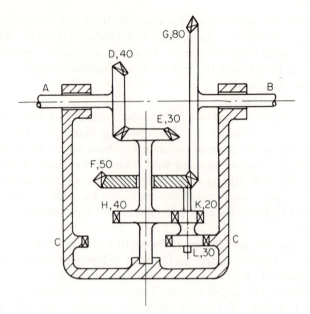

Figure 10.11

7. In the two-stage epicyclic shown in figure 10.12 the annulus I_2 is fixed and the arms C and D are attached to the shaft B. The annulus I_1 and sun wheel S_2 form a compound wheel which rotates freely about the axis AB. The numbers of teeth are S_1 24, S_2 28, I_1 66, I_2 62.

   If the input shaft A rotates at 3000 rev/min find the output speed B.
   [-589 rev/min]

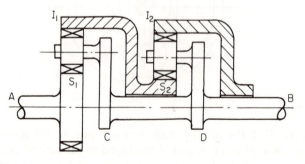

Figure 10.12

8. If in the previous problem the input power is 20 kW and the efficiency is 95 per cent, determine the torque on annulus I_2.
   [371 Nm]

9. The input shaft to a gearbox has an associated moment of inertia of 3 kg m², and the output shaft 20 kg m². If the input torque is 30 Nm

and there is a resisting torque of 50 Nm on the output shaft, find the value of gear ratio which gives maximum output acceleration, and determine this acceleration.

[4.74; 1.05 rad/s^2]

10. In the epicyclic gear shown in figure 10.13 the pinion A (30 teeth) drives wheel B which has 150 external and 120 internal teeth. Three planets C gear internally with B and externally with a fixed wheel D (60 teeth). The moments of inertia are: A 0.001 kg m^2, B 0.15 kg m^2, C 0.001 kg m^2, E 0.01 kg m^2. Each planet has a mass of 1 kg and is at a radial distance of 100 mm from the axis of D.

   Find the torque on A to accelerate E at 15 rad/s^2.

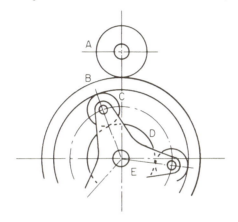

Figure 10.13

[0.92 Nm]

11. The engine of a motor car runs at 3500 rev/min when the road speed is 30 m/s. The mass of the car is 1000 kg of which the engine rotating parts are 10 kg at a radius of gyration of 0.15 m, and the road wheels are 100 kg at 0.25 m. The mechanical efficiency of the engine and transmission is 0.9, the wind and rolling resistance is 1000 N and the road wheel diameter is 0.75 m.

   Estimate the power developed by the engine when the car travels on a level road at 30 m/s with an acceleration of 1 m/s^2.

   [69 kW]

12. A racing-car engine can produce a constant torque of 1000 Nm at full throttle over a wide speed range. The axle ratio is 3.3:1 and the effective diameter of the road wheels is 0.75 m. The car has a mass of 900 kg, the moment of inertia of the engine is 2 kg m^2 and of each road wheel 3 kg m^2.

   When travelling at a steady 45 m/s the power absorbed is 60 kW. Assuming that windage and drag is proportional to the square of the speed, find the minimum time taken to accelerate from 25 m/s to 70 m/s.

   [7.2 s]

13. A pulley A, of 0.6 m diameter and moment of inertia 3 kg m^2, is connected by belting to a loose pulley B of 0.9 m diameter. When the speed of A is 50 rad/s the belt is suddenly shifted from B to a co-axial pulley C also of 0.9 m diameter, inertia 7.5 kg m^2, initially at rest.

Find the speeds when slipping has ceased, and the loss of kinetic energy.

If the belt tensions are 400 N and 200 N during slipping, find the time duration of slipping.

[23.7, 15.7 rad/s; 1980 Nm; 1.3 s]

14. A motorcycle and rider have a total mass of 200 kg. The moment of inertia of each wheel is 1.5 kg m^2, and of the engine rotating parts 0.12 kg m^2. The effective diameter of the road wheels is 0.6 m.

The motorcycle is travelling at 2 m/s when the rider changes into second gear with a reduction of 9:1. If the engine speed is 1200 rev/min immediately before the change, what effect will it have on the road speed?

[Increases to 2.68 m/s]

11 Kinetics of Machine Elements

11.1 INTRODUCTION

A machine has been defined as a combination of bodies used to transmit force and motion. The forces acting on the machine elements arise in a number of ways; for example, in a reciprocating engine there is the gas pressure on the piston and the load torque on the crankshaft, together with the reactions at the bearing surfaces. These forces can be analysed by the methods of statics if it is assumed that the mass of the engine parts can be neglected. However, even if the output speed is nominally constant there will be cyclical variations in the velocities of the piston and connecting-rod, and in consequence the force analysis should take account of mass–acceleration ('inertia') effects in these elements. Furthermore it is likely that the load will vary about its mean value, and this in turn will result in fluctuations in the crankshaft speed to an extent depending on the total kinetic energy in the system. It is these dynamic effects that will be considered in this chapter, generally in isolation from the effects of other external forces (including gravity).

11.2 INERTIA FORCES AND COUPLES

It was shown in chapter 3 that the acceleration of a body moving in a plane can be expressed in terms of the linear acceleration of the centre of gravity, a_G, together with the angular acceleration α. Applying D'Alembert's principle (section 4.3) equations 4.4 and 4.10 can be written

$$\Sigma F - ma_G = 0 \tag{11.1}$$

$$\Sigma M_G - I_G \alpha = 0 \tag{11.2}$$

where ΣF is the sum of the external forces and reactions acting on the body, and ΣM_G is the sum of the moments of these forces about the centre of gravity. $-ma_G$ and $-I_G\alpha$ are called the *inertia force* and the *inertia couple*,

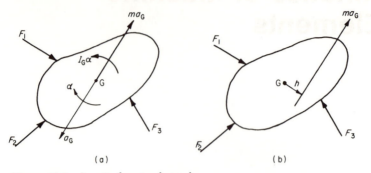

Figure 11.1 *Inertia force and couple*

and are imagined to act on the body in a sense opposite to the acceleration terms. The body is then treated as if it were in equilibrium under the action of the applied forces ΣF together with the inertia force and couple.

Figure 11.1a illustrates a typical force system, in which it is assumed that the magnitude and direction of the acceleration of the body have been determined. If desired, the inertia force and couple could be combined into a single force ma_G as in figure 11.1b, this force being displaced from the centre of gravity by a distance $h = I_G \alpha / ma_G$. In this example, if the directions of F_1, F_2 and F_3 were known, the conditions of equilibrium would enable their magnitudes to be determined.

Example 11.1

A single-cylinder petrol engine has a crank of 70 mm and a connecting-rod that is 250 mm between centres. The piston mass is 1.8 kg. The connecting-rod mass is 1.5 kg and its centre of gravity is 80 mm from the crank-pin centre; the radius of gyration about an axis through the centre of gravity is 100 mm.

Find the turning-moment on the crank due to the accelerations of the piston and connecting-rod when the crank is at 45 degrees after the top-dead-centre position and the engine is running at 1500 rev/min.

Figure 11.2a shows diagrammatically the configuration of the crank OA and the connecting-rod AP at the instant required (for convenience of layout the stroke has been drawn horizontally). R_p and R_c are the inertia forces due to the piston and connecting-rod respectively, N the normal reaction on the piston (friction will be negligible), and F_t, F_r the components of reaction on the big-end bearing of the connecting-rod.

The values of a_p, a_G and α are obtained by drawing the acceleration diagram, figure 11.2b

$$a_p = \mathbf{o'p'} = 1220 \text{ m/s}^2$$

$$a_G = \mathbf{o'g'} = 1470 \text{ m/s}^2$$

$$\alpha = \mathbf{n'p'}/CP = 4840 \text{ rad/s}^2$$

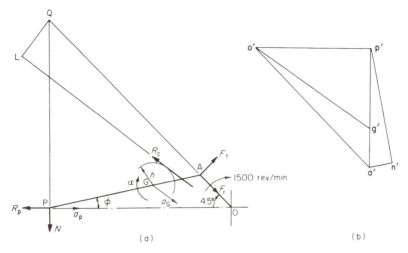

Figure 11.2

Then

$$R_{\mathrm{p}} = 1.8 \times 1220 = 2200 \text{ N}$$

$$R_{\mathrm{c}} = 1.5 \times 1470 = 2205 \text{ N}$$

$$I\alpha = 1.5 \times 0.1^2 \times 4840 = 72.6 \text{ Nm}$$

$$h = I\alpha/R_{\mathrm{c}} = 0.033 \text{ m}$$

By moments about Q

$$F_{\mathrm{t}} \times \mathrm{AQ} = R_{\mathrm{c}} \times \mathrm{LQ} + R_{\mathrm{p}} \times \mathrm{PQ}$$

$$F_{\mathrm{t}} = \frac{2205 \times 0.085 + 2200 \times 0.293}{0.345}$$

$$= 2410 \text{ N}$$

Turning-moment on the crank

$$= F_{\mathrm{t}} \times \mathrm{OA} = 169 \text{ Nm}$$

11.2.1   Equivalent Two-mass System

It is possible to replace any body of mass m by a dynamically equivalent system of two point masses which may be assumed to be joined by a rigid rod of zero mass, as in figure 11.3. The conditions to be satisfied are that the total mass must be the same, the centre of gravity G must be in the same place, and the moment of inertia I_{G} must be unchanged.

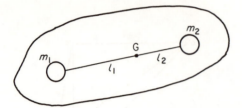

Figure 11.3   Equivalent two-mass system

Hence

$$m_1 + m_2 = m \tag{11.3}$$

$$m_1 l_1 = m_2 l_2 \tag{11.4}$$

$$m_1 l_1^2 + m_2 l_2^2 = I_G \tag{11.5}$$

These three equations contain four unknowns, and therefore one of these must be specified before the others can be determined. In practice it is usual to choose the position of one mass, say m_1, and then to calculate l_1, m_2 and l_2.

The two-mass system is particularly useful in simplifying the analysis of reciprocating-engine inertia effects, since it enables the connecting-rod to be replaced by point masses which may then be added to the piston and crank-pin respectively. It should be noted, however, that this will in general involve an approximation because it will not be possible to satisfy all three equations 11.3, 11.4 and 11.5, and at the same time specify the values of l_1 and l_2. Sufficient accuracy can usually be obtained by calculating m_1 and m_2 from equations 11.3 and 11.4 and accepting a small error in the moment of inertia.

11.2.2   Bending Effects due to Inertia Loading

Although the treatment used in preceding sections is adequate for determining the forces and moments transmitted through a machine, in the design of individual members it is necessary to take into account bending effects due to the distributed nature of the mass. Only the lateral component of acceleration is required, and a pin-jointed member is treated as a simply supported beam.

Example 11.2

In the engine of example 11.1 it may be assumed that the mass of the connecting-rod lying between its bearing centres is 1.2 kg and that it is uniformly distributed over the distance of 250 mm. For the position given find the maximum bending-moment in the connecting-rod at 1500 rev/min.

The connecting-rod makes an angle ϕ with the line of stroke OP in figure 11.2a, such that

$$\sin \phi = 70 \sin 45°/250$$

$$\phi = 11.4°$$

From figure 11.2b

$$a_p = 1220 \text{ m/s}^2 \text{ along PO}$$

$$a_a = 1730 \text{ m/s}^2 \text{ along AO}$$

When these are resolved at right angles to PA and multiplied by the mass distribution, the intensity of lateral inertia loading at P and A is obtained, that is

$$(1.2/0.25)\ 1220 \sin \phi = 1160 \text{ N/m at P}$$

$$(1.2/0.25)\ 1730 \sin(45° + \phi) = 6920 \text{ N/m at A}$$

Between P and A this intensity varies linearly as shown in figure 11.4, and this loading diagram will be divided into a rectangular and triangular portion for purposes of computation.

By moments about A, the lateral reaction at P is

$$\frac{1}{0.25}\left(1160 \times 0.25 \times \frac{0.25}{2} + \frac{5760}{2} \times 0.25 \times \frac{0.25}{3} \right)$$

$$= 385 \text{ N}$$

At a distance x m from P, the bending moment is

$$M = 385x - 1160x \times \frac{x}{2} - \frac{5760}{2} \times \frac{x}{0.25} \times x \times \frac{x}{3}$$

$$= 385x - 580x^2 - 3840x^3$$

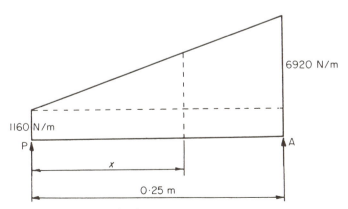

6920 N/m

1160 N/m

P

A

x

0·25 m

Figure 11.4   Inertia loading

The maximum bending-moment occurs at a position found by differentiating and equating to zero, thus

$$385 - 1160x - 11\,520x^2 = 0$$

Taking the positive root

$$x = 0.139 \text{ m}$$

Then, from above

$$M_{max} = 32 \text{ Nm}$$

11.3   FLUCTUATION OF ENERGY AND SPEED

A reciprocating engine, because of the variation in gas pressure and the obliquity of the connecting-rod, will always produce an uneven turning-moment on the crank. This cyclic fluctuation will be most marked in a single-cylinder four-stroke engine, which will have a turning-moment diagram of the type shown in figure 11.5. Clearly this could be made much more uniform by increasing the number of cylinders to four and arranging the firing strokes to be at 180° intervals.

However, even in multi-cylinder engines running against a constant load there will be some fluctuations of driving torque above and below the mean. This will lead to a cyclic fluctuation of speed as the energy in the system alternately increases and decreases. In order to limit the speed variation to an acceptable value it is generally necessary to add a flywheel to the system for the purpose of storing the energy during the periods of excess torque and releasing it during the periods of torque deficiency.

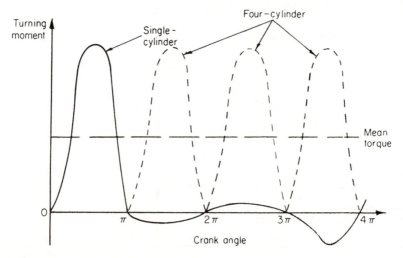

Figure 11.5   Turning-moment diagram

A similar situation arises when a constant-torque motor (for example, an electric or hydrostatic motor) drives a load such as a mechanical press or machine tool which requires a non-uniform operating torque. Again there will be a fluctuation of speed about the mean.

In either case, if T_D is the driving torque and T_L the load torque, the *fluctuation of energy* in the system during an angular rotation θ is

$$\Delta E = \int (T_D - T_L)\, d\theta \qquad (11.6)$$

$$= \tfrac{1}{2} I(\omega_1^2 - \omega_2^2) \qquad (11.7)$$

where I is the moment of inertia of the flywheel and rotating parts, and the angular velocity changes from ω_1 to ω_2. If the variation in T_D and T_L is known, then the problem is one of determining, usually by a semi-graphical method, the greatest value of ΔE during a complete cycle of operation. Corresponding to this the maximum overall *fluctuation of speed* is given by $\omega_{max} - \omega_{min}$ where these are the extreme values of ω_1 and ω_2. If this fluctuation is on either side of a mean speed ω_0, then ω_0 is approximately equal to $\tfrac{1}{2}(\omega_{max} + \omega_{min})$ and equation 11.7 may be written

$$\Delta E = \tfrac{1}{2} I(\omega_{max} - \omega_{min})(\omega_{max} + \omega_{min})$$

$$= I(\omega_{max} - \omega_{min})\omega_0 \qquad (11.8)$$

It should be noted that areas on the turning-moment diagram represent work or energy units, and also that, since the speed is increasing or decreasing at any instant according to whether T_D is greater or less than T_L, at the cross-over points (where $T_D = T_L$) the speed passes through a maximum or minimum value. It is therefore necessary to evaluate the 'intercepted' areas between T_D and T_L (which will alternate between positive and negative energy changes) and to use these to determine the points of greatest and least energy (and hence speed from equation 11.8).

Example 11.3

Figure 11.6 shows the turning-moment diagram for a multi-cylinder engine which is running at 800 rev/min against a constant load torque. Intercepted areas between the driving and load torques are, in sequence, -52, $+124$, -92, $+140$, -85, $+72$ and -107 mm^2, where the scales are 1 mm = 600 Nm vertically and 1 mm = $3°$ horizontally.

If the total fluctuation of speed is not to exceed 2 per cent of the mean, find the least radius of gyration of the flywheel if its mass is to be 500 kg.

Let E be the total energy in the flywheel at the beginning of the cycle. Then the values at the end of each intercepted area are as shown on figure 11.6.

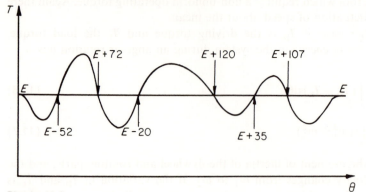

Figure 11.6

Hence

$$\Delta E = E_{\max} - E_{\min}$$

$$= E + 120 - (E - 52)$$

$$= 172 \text{ mm}^2$$

$$= 172 \times 600 \times \pi/60 = 5404 \text{ Nm}$$

From equation 11.8

$$5404 = 500k^2 \times \frac{2}{100}\left(\frac{800 \times 2\pi}{60}\right)^2$$

$$k = 0.277 \text{ m}$$

Example 11.4

A constant-torque motor runs at a mean speed of 300 rev/min and drives a machine that requires a torque increasing uniformly from 600 Nm to 2400 Nm while the shaft rotates through 45°, remains constant for the next 90°, decreases uniformly to 600 Nm in the next 45° and is constant for the remaining 180° of each revolution.

Find the output power of the motor and the total moment of inertia required to limit the fluctuation of speed to ±3 per cent of the mean.

The value of the constant driving torque must be equal to the mean value of the load torque in figure 11.7, that is

$$T_D = (\text{total area under load torque})/2\pi$$

$$= (600 \times 2\pi + 1800 \times \pi/4 + 1800 \times \pi/2)/2\pi = 1275 \text{ Nm}$$

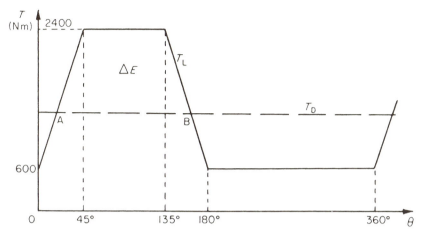

Figure 11.7

$$\text{Power} = T_D \omega_0$$

$$= 1275 \times 300 \times 2\pi/60 \text{ W}$$

$$= 40.1 \text{ kW}$$

The positions of the intersecting points A and B can be found by proportion, for example the value of θ at A is

$$\frac{1275 - 600}{2400 - 600} \times 45 = 16.875^\circ$$

$$= 0.2945 \text{ rad}$$

$$\Delta E = \text{area above } T_D \text{ between A and B}$$

$$= \tfrac{1}{2}(\pi - 2 \times 0.2945 + \pi/2)(2400 - 1275)$$

$$= 2319 \text{ Nm}$$

Equating this to $I(\omega_{max} - \omega_{min})\omega_0$ gives

$$I = \frac{2319 \times 100}{6} \left(\frac{60}{300 \times 2\pi} \right)^2$$

$$= 39.2 \text{ kg m}^2$$

Example 11.5

The crankshaft torque of a multi-cylinder engine is given by $T_D = 75 + 10 \sin 3\theta$ Nm, and the engine is coupled to a machine requiring a torque $T_L = 75 + 40 \sin \theta$ Nm. If the rotating parts have a total moment of inertia of 1.25 kg m^2 determine the maximum fluctuation of energy and the maximum angular acceleration.

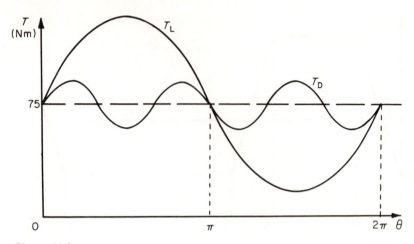

Figure 11.8

Figure 11.8 shows the variation of T_D and T_L with crank angle. It can be seen that the only cross-over points are at $\theta = 0$, π, 2π, hence from equation 11.6

$$\Delta E = \int_0^\pi (40 \sin \theta - 10 \sin 3\theta) \, d\theta$$

$$= 73.3 \text{ Nm}$$

The maximum positive acceleration occurs at $\theta = 3\pi/2$, where

$$\alpha = (T_D - T_L)/I$$

$$= 50/1.25 = 40 \text{ rad/s}^2$$

11.3 CAMS

Cam behaviour is another good example where the non-uniform motion of a machine element may lead to significant inertia effects. In most applications, uniform rotational motion of the cam shaft gives a non-uniform cyclic motion to the follower by means of a sliding pair. The valve operating gear in an internal combustion engine is one of the more common applications, and many different configurations are possible, figure 11.9.

The behaviour of the follower can be described by means of a *lift diagram*, such as that shown in figure 11.10. The important characteristics of the motion depend on the particular application, and may include

(a) Cam angle at commencement of lift and end of fall.
(b) Rise and fall time, which will be a function of cam angular velocity.

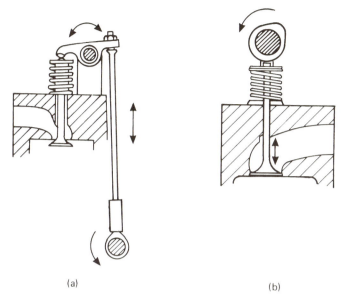

(a)

(b)

Figure 11.9 (a) Rocker type valve gear; (b) overhead camshaft

(c) Maximum lift.
(d) Dwell at maximum lift.
(e) Maximum acceleration, since this will determine the inertia forces.

In most cases compromises have to be made between the desired follower motion based on performance and acceptable cost.

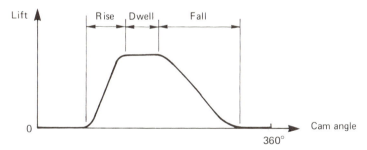

Figure 11.10   Cam lift diagram

11.4.1 Eccentric Circular Cam with Flat Follower

To illustrate some of the analysis techniques, a very simple cam mechanism will be considered, figure 11.11. The circular cam has a radius r and rotates about an axis O which is a distance e from the geometric centre C. The

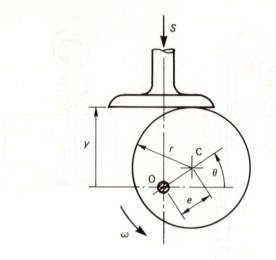

Figure 11.11

follower is constrained to move in a straight line through O, and is held in contact with the cam by a spring (not shown) that generates a force S. The position of the follower is

$$y = r + e \sin \theta \qquad (11.9)$$

By differentiation with respect to time

$$v = \dot{y} = e\omega \cos \theta \qquad (11.10)$$

and

$$a = \ddot{y} = -e\omega^2 \sin \theta \qquad (11.11)$$

and this is seen to be simple harmonic motion. (Kinematically, the system is equivalent to the Scotch yoke introduced in chapter 4.) There is no dwell at maximum or minimum lift in this case; the maximum acceleration is $e\omega^2$ and occurs when the follower is at its upper and lowermost positions, in the downward and upward directions respectively.

The motion described will only occur if the follower remains in contact with the cam, and this determines the characteristics of the spring. Thus

$$S \geqslant -ma$$

If the spring is assumed to have a constant stiffness k (such as given by a close coiled spring) then

$$S = k(y_0 + y - [r - e]) \geqslant -me\omega^2 \sin \theta$$

where y_0 is the axial displacement of the spring to give a *preload* when y is a minimum, that is, $(r - e)$. Thus

$$k \geqslant \frac{me\omega^2 \sin \theta}{y_o + e + e \sin \theta} \qquad \text{for all } \theta$$

and this is a maximum when $\sin \theta = 1$ giving

$$k \geqslant (me\omega^2)/(y_o + 2e) \tag{11.12}$$

If k does not satisfy equation 11.12 then the follower will 'bounce' on the cam. This is one of the possible factors which determine the maximum speed at which an internal combustion engine will run. Fitting stiffer valve springs is usually a relatively cheap method of improving engine performance, albeit at the cost of higher loads. Adequate preload is also necessary to ensure that the valve is held closed during the appropriate part of the cycle.

In figure 11.11 it is clear that sliding occurs between the lower face of the follower and the curved face of the cam. This will result in work done by friction forces with the consequent energy loss appearing as heat. Excessive sliding under load will also result in wear.

11.4.2 Eccentric Circular Cam with Curved Follower

It is clear from figure 11.11 that the follower must have a width of at least $2e$. By using a follower that has a curved face, figure 11.12, the width requirement is reduced, and, for the same cam angular velocity, so is the sliding velocity. Followers having a circular arc form are common, and if the radius can be made sufficiently small, the follower may be in the form of a roller, thus eliminating sliding altogether. A similar analysis to that for the

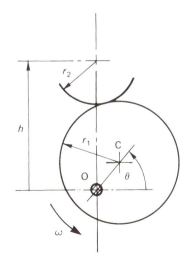

Figure 11.12

flat follower gives

$$y = e \sin \theta + (r_1 + r_2)\sqrt{[1 - e^2 \cos^2 \theta/(r_1 + r_2)^2]} \qquad (11.13]$$

$$v = e\omega \left[\cos \theta + \frac{e \sin 2\theta}{2\sqrt{[(r_1 + r_2)^2 - e^2 \cos^2 \theta]}} \right] \qquad (11.14)$$

$$a = e\omega^2 \left[-\sin \theta + \frac{e[(r_1 + r_2)^2 \cos 2\theta + e^2 \sin^4 \theta]}{[(r_1 + r_2)^2 - e^2 \sin^2 \theta]^{3/2}} \right] \qquad (11.15)$$

The influence of follower curvature is clearly seen in the second term of equation 11.15, which tends to zero as r_2 tends to infinity, and the follower becomes flat. When r_2 is zero, the follower is described as *knife edge*. The effect of these two extreme conditions on follower motion is shown in figure 11.13 for the particular case with $r = 10$ mm and $e = 5$ mm.

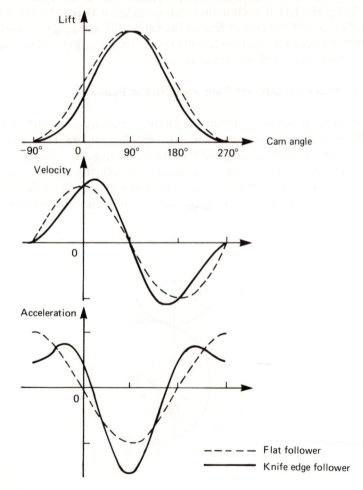

Figure 11.13

11.4.3   Other Cam Shapes

Most cams have rather more complex shapes than those considered in sections 11.4.1 and 11.4.2. Typically they have a circular arc *base* and an elongated curve, often with an axis of symmetry. Figure 11.14 shows a symmetrical example using circular arcs, which are easy to define. The follower rise occurs in two stages, first, while it is in contact with the flank of radius r_2, and second, while in contact with the nose of radius r_3.

Analysis follows exactly the same procedure as already described, but must be carried out separately for each different arc (each being considered as part of an eccentric circular cam), thus giving a series of functions for lift, velocity and acceleration. Although the cam profile appears smooth and continuous, the acceleration functions will show considerable discontinuities at the changes of curvature, and cams of this type would perform badly at all but the slowest speeds. In practice, the curves are blended to avoid these discontinuities, and maximum acceleration is often used as the criterion for design. Detailed analysis of such cams is generally only possible using numerical techniques.

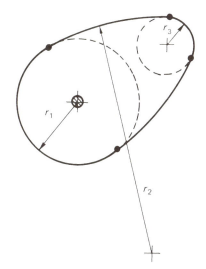

Figure 11.14

11.5   PROBLEMS

1. In the mechanism of figure 11.15 AB = 60 mm, BC = 160 mm, CD = 120 mm and AB is driven uniformly at 1200 rev/min. BC has a mass of 5 kg and a radius of gyration of 45 mm about its centre of mass, which

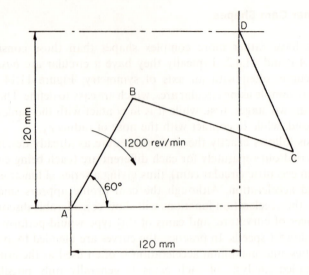

Figure 11.15

is 60 mm from B. Determine the reactions at the pins B and C due to the intertia of BC when AB is passing through the position shown. [3480, 470 N]

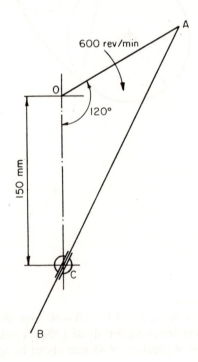

Figure 11.16

2. In the mechanism shown in figure 11.16 the crank OA is 100 mm long and rotates at 600 rev/min. The rod AB is 300 mm long and passes through a frictionless slider at C that can oscillate about an axis parallel to the crankshaft. AB is of uniform cross-section and has a mass of 6 kg. Find the turning-moment on the crankshaft to overcome the inertia of AB when the crank is at 120° to OC.
   [50.6 Nm]

3. An IC engine has a stroke of 110 mm and a bore of 80 mm. The connecting-rod is 180 mm between centres and has a total mass of 1.3 kg; its centre of mass is 130 mm from the small end centre and the radius of gyration about the mass centre is 75 mm. The reciprocating mass is 1.8 kg.

   Determine the magnitude of the resultant force on the crank-pin, neglecting friction and gravity, when the crank is 30° after the top-dead-centre position and rotating at 1600 rev/min, if the gas pressure on the piston is 1.9 N/mm².
   [5210 N]

4. A rigid uniform bar AB is 0.5 m long and has a mass of 6 kg. The bar is pin-jointed at A and at a point C on AB distant 0.35 m from A, to blocks that are constrained to move along two straight lines at right angles to one another. Find the bending moment at C due to inertia effects when the end A is moving at a constant speed of 7.5 m/s and the bar is inclined at 30° to the path of A.
   [193 Nm]

5. The torque exerted on the crankshaft of an engine is given in Nm by the expression $10\,500 + 1620 \sin 2\theta - 1340 \cos 2\theta$, where θ is the crank angle measured from inner dead-centre. Assuming the resisting torque to be constant, determine (a) the power of the engine at a mean speed of 1500 rev/min, (b) the moment of inertia of the flywheel if the speed variation is not to exceed ±0.5 per cent, and (c) the angular acceleration of the flywheel when the crank has turned through 30° from inner dead-centre.
   [1650 kW, 8.51 kg m², 86 rad/s²]

6. The turning-moment diagram for one revolution of a multi-cylinder engine shows the following intercepted areas in mm² above and below the load torque line: $-5, +63, -42, +51, -49, +36, -58, +42, -38$. The vertical and horizontal scales are 1 mm equal to 1000 Nm and 5° respectively. If the mean speed is 500 rev/min and the overall fluctuation of speed is not to exceed 1.5 per cent, determine the moment of inertia of the flywheel.
   [152 kg m²]

7. A six-cylinder four-stroke engine has cranks spaced at 120° to each other, and the turning-moment diagram for each power stroke may be approximated to a triangle based on 180°, with a maximum value of

1500 Nm after 60°. Torque during the other three strokes for each cylinder may be neglected, and the firing order is such that the power strokes are at regular intervals. Find the mean torque of the engine and the flywheel inertia to keep the speed within 1800 ± 30 rev/min.

[1125 Nm; 0.166 kg m^2]

8. The cycle of operations performed by a machine extends over 3 revolutions. The torque required has a constant value of 400 Nm for one revolution, zero for the next revolution, 550 Nm for the first half of the third revolution and zero for the second half. If the driving torque is constant, the mean speed is 180 rev/min and the flywheel has a mass of 500 kg at a radius of gyration of 0.5 m, calculate (a) the power required, (b) the percentage fluctuation of speed, and (c) the greatest acceleration and retardation.

[4.24 kW; 3.18 per cent; 1.8, 2.6 rad/s^2]

9. If the turning-moment diagram of an engine can be written $14\,000 + 4665 \sin 3\theta$ Nm and the load torque $14\,000 + 2000 \sin \theta$ Nm, calculate the fluctuation of energy and speed if the moment of inertia of the flywheel is 100 kg m^2 and the mean speed is 600 rev/min.

[5310 Nm, 8.07 rev/min]

10. An eccentric circular cam operates a flat-footed follower of mass 420 g at 40 Hz to give a lift of 18 mm against a linear return spring. If the normal reaction between the follower and the cam is not to fall below 35 N, what stiffness of spring should be used?

[11.3 N/mm]

11. In an engine valve gear, such as shown in figure 11.9a, the valve has a mass of 0.11 kg, the pushrod has a mass of 0.15 kg and the rocker has a moment of inertia of 87×10^{-8} kg m^2 about its pivot, which is midway between the contact points with the valve and pushrod. The effective length of the rocker is 76 mm. The cam has a base circle of radius 19 mm, a nose of radius 3.8 mm and a flank of radius 100 mm, and moves the valve through a distance of 10 mm. If the spring has a preload of 85 N and a stiffness of 46 N/mm, what is the maximum speed at which the system will operate satisfactorily?

[2483 rev/min]

12. The valve in a diesel engine is open while the cam shaft rotates through 110°, and is fully open for 14° at its maximum lift of 9 mm. The opening and closing times are equal, and the profile is to be such that the maximum acceleration and deceleration are minimised. Sketch the lift, velocity and acceleration diagrams for the valve as functions of camshaft angle, and determine the value of the maximum acceleration when the camshaft rotates at 900 rev/min.

[457 m/s^2]

12 Balancing of Machines

12.1 INTRODUCTION

The inertia forces and couples produced by the accelerations of machine members have been discussed in chapter 11 for a wide variety of mechanisms. It should be clear that in the majority of machines such forces cannot be avoided and must be taken into account when designing the members, the bearing surfaces and the foundations. The severity of the inertia loading increases as the square of the running speed, so that for high-speed machinery the effects can be very objectionable, often resulting in vibrations, noise emission and premature failure of a fatigue nature.

This chapter will consider the effects, and the alleviation, of inertia forces in two particular cases that are of major practical importance—rotating members and reciprocating engines. Some of the principles developed could be used for balancing members in more complex machinery, but each case would have to be dealt with in an empirical fashion.

12.2 BALANCING OF ROTATING MASSES

If a mass m is attached to a shaft rotating with angular velocity ω about a fixed axis and the centre of mass is at a distance r from the axis, then the inertia force on the shaft is

$$F = mr\omega^2 \tag{12.1}$$

in a radially outwards direction with respect to r (note that F is constant in magnitude but rotating in direction).

Where a number of such masses are attached to the same shaft (as in figure 12.1a) the resultant inertia force can be obtained from the vector sum (shown in figure 12.1b), taking into account the magnitude and relative directions of the individual forces (a graphical solution is convenient, and for this purpose the constant factor ω^2 can be omitted).

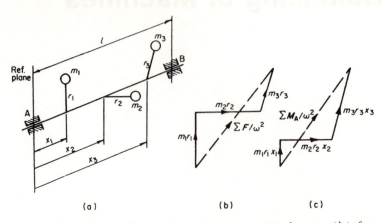

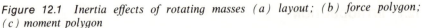

Figure 12.1   Inertia effects of rotating masses (a) layout; (b) force polygon; (c) moment polygon

If the masses are attached at different points along the shaft then they will rotate in parallel planes and this will give rise to moment effects. The magnitudes of these moments will depend on the position chosen for the 'reference plane'. In the illustration shown, suppose the shaft is supported in two bearings, A and B, the left-hand bearing being taken as the reference plane. The resultant moment about this reference plane can be written

$$\Sigma M_A = \Sigma mrx\omega^2 \tag{12.2}$$

where x is the distance measured along the axis and may take positive or negative values (if there are masses on both sides of the reference plane). Because all the individual moments are in planes through the shaft axis they would normally be represented by vectors perpendicular to this axis and to the direction of r. It is acceptable however, and greatly simplifies the moment diagram, to rotate all vectors through 90° about the shaft axis so that they fall along the direction of r (as in figure 12.1c). The out-of-balance moment derived from equation 12.2 and figure 12.1c will give rise to bearing reaction R_B at the right-hand end such that

$$R_B = \Sigma M_A/l \tag{12.3}$$

Finally, the bearing reaction R_A at the left-hand end can be obtained from figure 12.1b and the relation

$$\Sigma F = R_A + R_B \tag{12.4}$$

It is seen that in the general case of rotating masses there will be an unbalanced force and an unbalanced moment. For complete balance of the inertia effects (so that no forces are transmitted to the bearings) both resultant force and resultant moment must be zero (that is, the force and moment polygons must close). These two conditions can usefully be considered separately, zero force being referred to as 'static balance', and zero moment (about any and all reference planes) as 'dynamic balance'.

Static balance

It can easily be verified that if $\Sigma F = 0$ (that is, $\Sigma mr = 0$) the shaft will be in equilibrium in all positions when at rest under the effects of gravity. Hence the condition of static balance indicates that the resultant rotating inertia force is zero. Since this is a necessary condition for complete balance (though not a sufficient condition unless all the masses are in the same plane) and can be checked by simple mounting of the shaft on frictionless bearings, it is of some practical significance.

Example 12.1

A shaft 2 m long is supported in bearings at 200 mm from each end and carries three pulleys, one at each end and one at the mid-point. The pulleys are out-of-balance to the extent of 0.06, 0.08, and 0.09 kg m in order from one end, but are keyed to the shaft so as to achieve static balance. Find (a) the relative angular settings of the three pulleys, and (b) the dynamic load on each bearing when the shaft rotates at 720 rev/min.

The mr products for the three pulleys are 0.06, 0.08 and 0.09 kg m, and for static balance these form a closed triangle, as in figure 12.2a, in which the 0.06 has been drawn at the instant when it is in a horizontal direction. Transferring the vectors on to the radial diagram of figure 12.2b gives the relative directions of the out-of-balance forces, that is, 259° and 119° relative to the pulley with 0.06 out-of-balance (it may be thought that there are alternative solutions, but in fact they are only mirror images of the same solution and do not affect the second part of the problem).

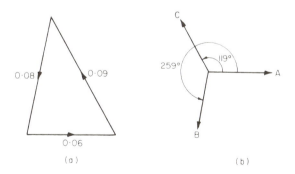

(a)   (b)

Figure 12.2

The axial dimensions of the system are shown in figure 12.3a, in which the reference plane has been taken through the left-hand bearing. The mrx products for the pulleys are then -0.012, 0.064, and 0.162 kg m². These are

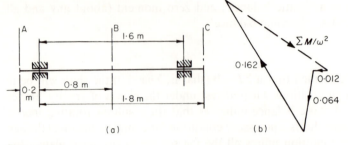

Figure 12.3

added vectorially in figure 12.3b, noting that the negative moment vector is drawn in the opposite direction to the radial force. From this

$$\Sigma M = 0.129\omega^2$$

and the right-hand bearing reaction

$$R_r = \Sigma M / 1.6$$

$$= 0.129 \left(\frac{720 \times 2\pi}{60} \right)^2 \frac{1}{1.6}$$

$$= 460 \text{ N}$$

Since there is no net unbalanced force in the system, the left-hand bearing reaction R_1 is equal and opposite to R_r. In fact ΣM is a pure couple and will have the same value wherever the reference plane is taken.

Dynamic Balance

If a rotating system is in static balance (that is, no resultant radial inertia force) there may still be an unbalanced couple if individual forces arise in different planes. Dynamic balance will only be achieved if this couple also is zero, that is $\Sigma M = 0$.

For complete balance it is therefore necessary that both the force polygon and the moment polygon close, and provided this is so it is immaterial where the reference plane is taken (the reader should also satisfy himself that a system is in complete balance if $\Sigma M = 0$ about any two different reference planes, but this is usually a more devious approach to any problem). Since the closure of each polygon effectively satisfies two conditions of equilibrium (zero net component in two directions) complete balance requires the selection of values for four quantities. These may be any combination of (a) the magnitude of individual mr products, (b) the position x along the shaft, and (c) the relative radial disposition θ.

In a numerical problem it is almost always necessary to draw the moment polygon first, since this permits the elimination of one of the 'unknowns' by choosing the reference plane to pass through it.

Example 12.2

A, B, C and D are four masses attached to a rotating shaft, with their centres of mass lying at radii of 100, 125, 200 and 150 mm respectively. The planes in which the masses rotate are spaced 0.6 m apart and the magnitudes of A, C and D are 15, 10 and 8 kg respectively. Find the value of the mass B and the relative angular settings for the shaft to be in complete balance.

It is convenient to set up a table of values of the given quantities and derived products. From this the four unknowns are clearly identified and it will be seen that the moment polygon can be reduced to a triangle with sides of known length by taking the reference plane through B.

Plane	m (kg)	r (mm)	x (m)	mrx (kg m^2)	mr (kg m)	θ (°)
A	15	100	−0.6	−0.9	1.5	0
B		125	0	0		
C	10	200	0.6	1.2	2.0	
D	8	150	1.2	1.44	1.2	

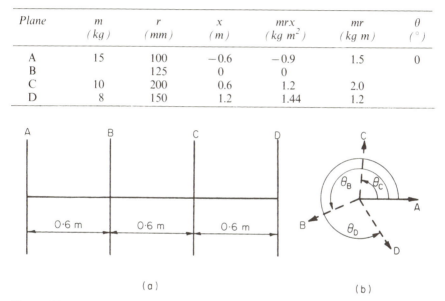

(a)   (b)

Figure 12.4

Figure 12.4a shows the axial dispositions of the masses, and figure 12.4b the radial positions (as determined later) relative to A which is taken as the reference direction.

The angles for C and D are determined from the moment polygon (figure 12.5a), in which the vector for A is negative. Care must be taken in transferring these vectors to the radial diagram so that positive vectors correspond to the outwards radial direction. Values of θ should then be measured on the radial diagram, not on the vector polygon. In this case

$$\theta_C = 85° \qquad \theta_D = 304°$$

The magnitude and radial direction for the mass B can now be found

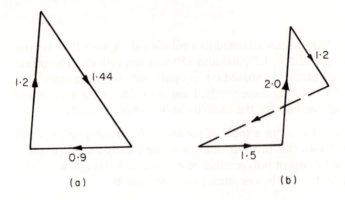

Figure 12.5   (a) Moment polygon; (b) force polygon

by completing the force polygon (figure 12.5b), the closing line giving $(mr)_B = 2.54$ kg m^2, from which

$$m_B = 20.3 \text{ kg}, \qquad \theta_B = 203°$$

12.3   FORCES IN A RECIPROCATING ENGINE

It was shown in section 7.3.1 that the acceleration of the reciprocating parts (that is, the piston) of an engine is approximately $r\omega^2 \cos\theta + (r^2\omega^2/l) \cos 2\theta$ at a crankshaft speed ω and a crank angle θ from the dead-centre position. This gives rise to an inertia force equal to

$$mr\omega^2 \left(\cos\theta + \frac{1}{n} \cos 2\theta \right) \qquad (12.5)$$

where $n = l/r$ is the connecting-rod:crank ratio and m is the mass of the reciprocating parts (piston together with a proportion of the connecting-rod mass as explained in section 11.2.1).

Figure 12.6a shows the moving parts of the engine acted on by the gas force P and load torque T together with the reciprocating and rotating inertia forces and the reactions at the bearing surfaces, neglecting friction (m_{rot} is the effective mass of the crank-pin together with a proportion of the connecting-rod). Figure 12.6b shows the forces transmitted to the engine block, and it will be seen that, apart from the reaction torque T produced by the forces N, there are unbalanced forces transmitted through the bearings equal to the reciprocating and rotating inertia forces.

The balancing of the rotating mass is usually achieved by adding mass

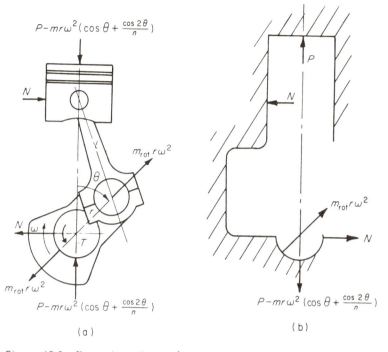

Figure 12.6   *Forces in engine mechanism*

to the crank web diametrically opposite the crank-pin. This has the advantage of balancing the force in planes as near as possible to the plane in which it arises, so reducing the stress levels in the crankshaft. However, it involves increasing the mass and moment of inertia of the crankshaft, and this may have undesirable side effects not only in increased overall engine weight but also on lower torsional resonances (see section 14.6). An alternative solution is often available for multi-cylinder engines when, due to the symmetry of the layout, the rotating masses balance as a group. A compromise must usually be made between reduced bending stress in the crankshaft and minimum overall weight.

The remainder of this chapter is devoted to considering the balance of the reciprocating inertia as expressed in equation 12.5, and for this purpose the two terms are separated and called *primary force* ($mr\omega^2 \cos\theta$) and *secondary force* (($mr\omega^2/n)\cos 2\theta$). Each of these is a force that acts along the line of stroke and varies harmonically in magnitude, the primary force at a frequency corresponding to the crank speed and the secondary force at twice the crank speed. There are also higher harmonics of piston acceleration (of fourth, sixth and even orders only) but successive coefficients are so small in relation to the primary and secondary terms that they can be neglected except at high speeds in engines that are balanced for primary and secondary effects.

12.4 BALANCING OF A SINGLE-CYLINDER ENGINE

By comparison with the effects of rotating inertia it can readily be seen that the primary force is equal to the component along the line of stroke of the radial force that would be produced by a rotating mass m attached at the crank-pin. This seems to suggest that the primary force could be balanced by an equal mass attached diametrically opposite to the crank-pin. However, although forces along the line of stroke are then balanced, this balance is achieved at the expense of introducing an out-of-balance of equal magnitude perpendicular to the stroke. In fact, a reciprocating inertia force cannot be completely eliminated by a single rotating mass. The optimum partial balance that can be obtained is by adding at a radius r a rotating mass equal to half the reciprocating mass. The resulting imbalance is then reduced to $\frac{1}{2}mr\omega^2$, this force varying in direction as the engine rotates.

If, however, a second rotating mass is added and caused to rotate in a sense opposite to the crank rotation, it is possible to balance the primary force along the line of stroke and at the same time to avoid introducing unbalanced forces at right-angles to the stroke. This contra-rotating balance system is shown in figure 12.7a. A similar system for balancing the secondary force is shown in figure 12.7b, but here a further complication arises because the balance masses have to be driven at twice the crank speed (and neither

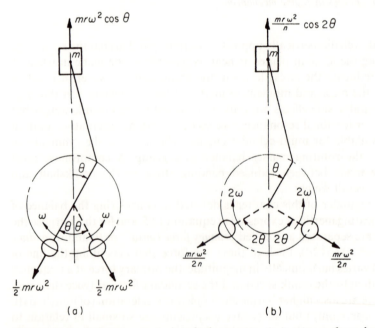

Figure 12.7 Balancing a single-cylinder engine (a) primary-force balance; (b) secondary-force balance

of them can be physically attached to the crank). Because of the added complexity and increased mass this method of complete balance is rarely adopted for a single-cylinder engine, but the principle has been used to balance the secondary forces of four-cylinder engines, and in this form it is known as the Lanchester balancer.

12.5 BALANCING OF MULTI-CYLINDER ENGINES

The majority of reciprocating engines consist of a number of cylinders with pistons linked to the same crankshaft and lines of stroke that are parallel. These are called *in-line* engines. If the strokes are not all in the same direction the configuration will result in either a V-engine or a radial engine

12.5.1 In-line Engines

With more than one cylinder it will be possible to obtain a degree of balance of the reciprocating inertia forces by suitable choice of relative crank spacing (both radially and axially). The objective is often made easier by the axial symmetry of the engine and by the requirement for equal firing intervals that results in even radial spacing between cranks. As with rotating balance there will be moment effects obtained by multiplying the primary and secondary forces by the distance x between each cylinder and a reference plane. For symmetrical engines it is normal practice to take the central plane as reference plane. Then, for complete balance, omitting the factor ω^2

$$\Sigma mr \cos \theta = 0 \tag{12.6}$$

$$\Sigma (mr/n) \cos 2\theta = 0 \tag{12.7}$$

$$\Sigma mrx \cos \theta = 0 \tag{12.8}$$

$$\Sigma (mrx/n) \cos 2\theta = 0 \tag{12.9}$$

Each of these summations can be represented by vectors of the appropriate lengths and inclined at corresponding θ values to the line of stroke for the primary forces and moments (equations 12.6 and 12.8), and at 2θ for the secondary forces and moments (equations 12.7 and 12.9). The *instantaneous* value of each term is then given by the projection of the vector on to the line of stroke, but usually it is only necessary to check whether each vector polygon closes or to determine the *maximum* resultant out-of-balance (which will be given by the length of line required for closure).

Example 12.3

An air compressor has four vertical cylinders 1, 2, 3, 4 in line, and the driving cranks are at 90° intervals and reach their top-dead-centre in that order. The cranks are 150 mm radius, the connecting-rods are 500 mm long, and the cylinder centre-lines are 400 mm apart. The reciprocating parts for each cylinder are 20 kg and the speed of rotation is 400 rev/min. Show that there are no out-of-balance primary or secondary forces and determine the out-of-balance couples and the positions of crank 1 at the instant when the maximum values occur.

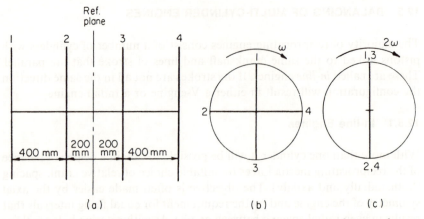

Figure 12.8 (*a*) *Cylinder spacing;* (*b*) *primary cranks;* (*c*) *secondary cranks*

Figures 12.8a and b show the cylinder spacing and the primary (that is, actual) crank positions, at the instant when crank 1 is on top-dead-centre. Figure 12.8c shows the 'secondary' crank positions, in which all angles relative to the top-dead-centre have been doubled. The table below gives the values to be used for primary forces and moments (and since the connecting-rod:crank ratio is the same for all cylinders, these values can be divided by 3.33 and used to represent secondary effects).

Crank	mr (kg m)	x (m)	mrx (kg m^2)	θ (°)	2θ (°)
1	3	−0.6	−1.8	0	0
2	3	−0.2	−0.6	270	180
3	3	0.2	0.6	180	0
4	3	0.6	1.8	90	180

Figure 12.9 indicates that primary and secondary forces are in balance, a fact that follows directly from the equal values at each cylinder combined with the crank spacings.

Taking into account that cylinders 1 and 2 produce 'negative' moments, the primary-moment polygon appears as figure 12.10a and the maximum ouf-of-balance couple is given by the closing line, that is

$$(1.8 + 0.6)\sqrt{2}(400 \times 2\pi/60)^2 = 5955 \text{ Nm}$$

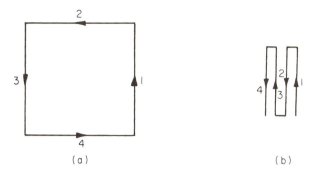

Figure 12.9   *(a) Primary forces; (b) secondary forces*

occurring when the dotted line falls along the line of stroke, that is when crank 1 has rotated 45° or 225° from top-dead-centre.

Similarly the maximum secondary out-of-balance couple is obtained from the polygon of figure 12.10b, and is equal to

$$\left(\frac{1.8 - 0.6 - 0.6 + 1.8}{3.33}\right)\left(\frac{400 \times 2\pi}{60}\right)^2 = 1263 \text{ Nm}$$

and this occurs when crank 1 is at 0°, 90°, 180° or 270°, since the secondary polygon rotates at 2ω.

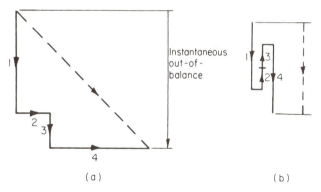

Figure 12.10   *(a) Primary moments; (b) secondary moments*

The reader should note that it is not possible to add directly the maximum primary and secondary effects, although there will be a crank position (in this case approximately 36° for crank 1) at which the combined instantaneous value of moment is a maximum. Since the forces are balanced, this moment is in fact a pure couple, independent of the position of the reference plane. An alternative crank layout, which would balance secondary moments also, would be obtained by changing the sequence to 1, 3, 4, 2, though in doing this the primary couple is increased by about 12 per cent.

Example 12.4

Show that a six-cylinder in-line four-stroke internal combustion engine that has equal reciprocating masses and crank spacing is in complete balance for a firing order 1–5–3–6–2–4.

For a four-stroke cycle each cylinder fires once in two revolutions of the crankshaft. Hence for equal firing intervals and six cylinders, one crank must be on top-dead-centre every 120°. With a firing order 1–5–3–6–2–4 this requires the crank angles to be as figure 12.11b and the secondary angles as figure 12.11c. It is clear from these radial diagrams that the primary and secondary forces will balance at all times (it is implicit that crank and connecting-rod dimensions are identical).

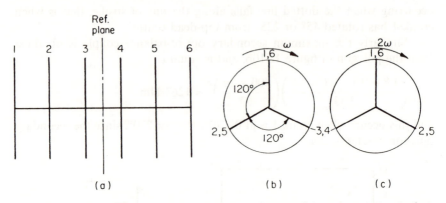

Figure 12.11   Six-cylinder four-stroke engine (a) layout; (b) primary; (c) secondary

Taking the central plane as reference it is easily shown that the primary and secondary moments balance (as laid out in figures 12.12a and b). This could also be inferred from the radial diagrams, which show that 1 balances 6, 2 balances 5, and 3 balances 4.

12.5.2   Balancing of Vee-engines

In practice, more engine torque and power is achieved by increasing capacity, but merely increasing the size of the cylinders is not generally favoured beyond a certain point because of the large out-of-balance effects. Increasing the number of cylinders gives more scope for balancing and also helps to smooth out the torque cycle, but multi-cylinder in-line engines start to become very long, and, in particular, the long crankshaft becomes less rigid, giving rise to vibration. A Vee-layout allows the engine to be made shorter and tends to avoid this problem.

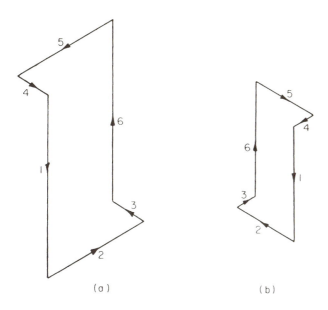

Figure 12.12   (a) Primary moments, (b) secondary moments

Figure 12.13 shows the simplest possible example of a Vee-engine, comprising only two slider–crank mechanisms, with the lines of stroke A and B symmetrically disposed about a vertical centre-line. If a four-stroke cycle with equally spaced working strokes (or uniform firing intervals) is used, the arrangement will be as shown in figure 12.13a, the two pistons always being at the same relative position along their respective lines of stroke, that is, both reach their dead-centre positions at the same instant. Alternatively, the connecting rods can be journalled on a common crank pin, as shown in figure 12.13b. This latter arrangement simplifies crankshaft design and manufacture, and gives further scope for keeping overall engine length small, but the firing is no longer uniform.

Most Vee-engines have a number of cylinders, usually at least three, in each *bank* (A and B), and arrangements with uniform firing and common crank pin are both in current use. In most cases the firing order is chosen so that consecutive working strokes occur on alternate banks.

In very simple arrangements, especially with uniform firing, out-of-balance effects can be determined by drawing vector diagrams in a very similar manner to that described for in-line engines. More generally, and especially where common crank pins are used, it is usually easier to resolve the inertia force associated with each cylinder onto the vertical and horizontal axes, Y and X, and to express the out-of-balance as separate summations in these two directions.

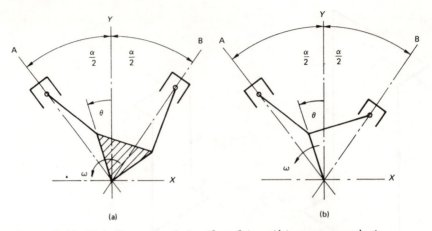

Figure 12.13 *Twin Vee-engine (a) uniform firing; (b) common crank pin*

Example 12.5

Compare the state of balance of the two systems shown in figure 12.13.

(a) *Uniform firing* (figure 12.13a)

In this case, as stated earlier, both pistons are at the same relative positions on their respective lines of stroke, and thus will generate maximum primary forces simultaneously when $\theta = \alpha/2$, that is, at their respective top dead centres. Figure 12.14 shows the components of the two primary forces, and it is clear that the horizontal components sum to zero while the vertical components give a resultant of $2mr\omega^2 \cos(\alpha/2)$, where the symbols are as

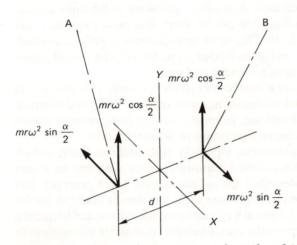

Figure 12.14 *Primary force components for uniform firing*

previously defined. It is also clear that there is a maximum primary couple of $mr\omega^2 d \sin(\alpha/2)$ about the Y-axis.

The maximum secondary forces will occur at the same position and give a vertical resultant of $(2mr\omega^2/n)\cos(\alpha/2)$, and a couple about the Y-axis of $(mr\omega^2 d/n)\sin(\alpha/2)$.

(b) *Common crank pin* (figure 12.13b)

In this arrangement the two pistons will not generate their maximum forces at the same time, and hence a more general approach is required.

The primary forces associated with the pistons are

$$PF_A = mr\omega^2 \cos\left(\frac{\alpha}{2} - \theta\right) \quad \text{and} \quad PF_B = mr\omega^2 \cos\left(\frac{\alpha}{2} + \theta\right)$$

Resolving these onto the Y-axis and adding gives the resultant primary force along the Y-axis

$$PF_Y = mr\omega^2 \cos(\alpha/2)\left[\cos\left(\frac{\alpha}{2} - \theta\right) + \cos\left(\frac{\alpha}{2} + \theta\right)\right]$$

$$= 2mr\omega^2 \cos^2(\alpha/2)\cos\theta$$

In a similar manner

$$PF_X = mr\omega^2 \sin(\alpha/2)\left[\cos\left(\frac{\alpha}{2} - \theta\right) - \cos\left(\frac{\alpha}{2} + \theta\right)\right]$$

$$= 2mr\omega^2 \sin^2(\alpha/2)\sin\theta$$

There is no convenient general expression for the total primary out-of-balance force, but note that if the Vee angle α is $90°$, then the two components reduce to

$$PF_Y = mr\omega^2 \cos\theta \quad \text{and} \quad PF_X = mr\omega^2 \sin\theta$$

These are the components of a constant force rotating with the crank, and are thus easily balanced by a single rotating mass of equivalent mr value opposite the crank. A general treatment for the primary moment is rather more difficult, but careful distribution of the balancing mass, and the reduction in cylinder centre distance resulting from the use of a common crank pin, will tend to keep the moment small.

The secondary forces can be treated in a similar manner. Thus

$$SF_A = \frac{mr\omega^2}{n}\cos^2\left(\frac{\alpha}{2} - \theta\right) \quad \text{and} \quad SF_B = \frac{mr\omega^2}{n}\cos^2\left(\frac{\alpha}{2} + \theta\right)$$

which, when resolved and summed along the X- and Y-axes give

$$SF_Y = \frac{2mr\omega^2}{n} \cos\frac{\alpha}{2} \cos\alpha \cos 2\theta$$

$$SF_X = \frac{2mr\omega^2}{n} \sin\frac{\alpha}{2} \sin\alpha \sin 2\theta$$

Again, it is not helpful to attempt to combine these components, but note that if $\alpha = 90°$, the Y-component of the secondary force is zero, and the X-component is $(\sqrt{2})(mr\omega^2/n) \sin 2\theta$.

SUMMARY

Rotating out of balance

For a single mass $F = mr\omega^2$
along a line joining the mass to the axis
and rotating with the shaft

For several masses $R = $ vector sum $F_1, F_2 \ldots$

and $M = $ vector sum of $F_1 d_1, F_2 d_2 \ldots$

Reciprocating out of balance

For a single piston

Primary force $PF = mr\omega^2 \cos\theta$
along the line of stroke and varying
harmonically at crank speed

Secondary force $SF = (mr\omega^2/n) \cos 2\theta$
along the line of stroke and varying
harmonically at twice crank speed

Balancing of multi-cylinder in-line and Vee-engines
by appropriate vector summation

12.6   PROBLEMS

1. A rotor mounted between centres for machining is given static balance by two masses bolted on near the ends. The rotor mass is 200 kg and the balance masses are 7.2 kg and 10 kg at radii of 350 mm and 250 mm respectively. The balance masses are at 135° to each other and at 375 mm on either side of the rotor mass centre.

   Determine the eccentricity of the mass centre of the unbalanced couple when rotating at 300 rev/min.
   [9.6 mm; 1716 Nm]

2. A shaft turning at 200 rev/min carries two uniform discs A and B of masses 10 kg and 8 kg respectively. The mass centres of the discs are each 2.5 mm from the axis of rotation and in mutually perpendicular directions.

The shaft is carried in bearings at C and D between A and B such that $AC = 0.2$ m, $AD = 0.6$ m, $AB = 0.8$ m. It is required to make the dynamic loading on the bearings equal and a minimum by adding a mass at a radius of 25 mm in a plane E.

Determine (a) the magnitude and the angular position of the mass in plane E relative to the mass centre of disc A, (b) the distance of plane E from plane A, and (c) the dynamic load on each bearing.
[0.28 kg at 141°; 1.42 m; 6.86 N]

3. Figure 12.15 shows a vertical spindle supported on a pivot bearing E at its lower end and by a plain journal bearing at D. The spindle carries three uniform discs A, B and C, each 100 mm diameter and 6.25 mm thick, made of material with a density of 7760 kg/m³. The disc A has a hole 18.75 mm diameter with its centre at 25 mm from the axis of the spindle. The disc B has a hole 25 mm diameter with centre 18.75 mm from the axis. The relative positions of these holes are 120°.

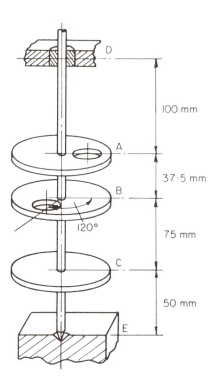

Figure 12.15

Find the position of a third hole, 12.5 mm diameter, to be drilled in C so that there will be no lateral thrust on bearing E when the spindle rotates. What will then be the side force on bearing D at a speed of 1800 rev/min?

[267° from A, 42.1 mm from axis; 5.96 N]

4. Three rotating masses, A = 14 kg, B = 11 kg, and C = 21 kg, are attached to a shaft with their centres of gravity at 55 mm, 80 mm and 30 mm respectively from the shaft axis. The angular positions of B and C from A are respectively 60° and 130° measured in the same direction. The distance between the planes of rotation of A and B is 1.5 m and between B and C is 2.5 m in the same direction.

Determine the magnitudes and angular positions with respect to A of two balance masses, each with its centre of gravity at 45 mm from the shaft axis, to be attached in planes midway between A and B and between B and C.

[35.2 kg at 186°; 29.4 kg at 306°]

5. A vertical reciprocating engine has three cylinders in line, A, B and C. The centre-lines of A and B are 1 m apart, and of B and C 1.2 m apart. The reciprocating parts are 100 kg for A, 110 kg for B, and 130 kg for C. The stroke is 750 mm and the connecting-rods are each 1375 mm, and the engine runs at 150 rev/min. Find the crank angles relative to A for balance of primary forces. What are then the maximum unbalanced primary couple and secondary force, and at what positions of crank A to they occur?

[B = 104°, C = 235°; 21 100 Nm, A at 34° or 214°; 2100 N, A at 61°, 151°, 241° or 331°]

6. The firing order for a four-cylinder in-line engine is to be 1–3–4–2 at equal crankshaft intervals. If the reciprocating mass and dimensions are the same for all cylinders and the cylinder spacing is d, investigate the degree of balance for (a) a two-stroke engine, (b) a four-stroke engine.

[(a) unbalanced primary couple, maximum value $\sqrt{10}mrd\omega^2$; (b) unbalanced secondary force, maximum value $4mr\omega^2/n$]

7. Figure 12.16 shows the layout of a four-crank 'symmetrical' engine in which the reciprocating mass is m_1 for cylinders 1 and 4 and m_2 for

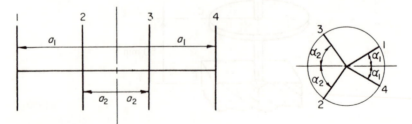

Figure 12.16

cylinders 2 and 3. Show that the arrangement is balanced for primary forces and couples and for secondary forces if

$$\frac{\cos\alpha_1}{\cos\alpha_2}=\frac{m_2}{m_1}, \qquad \frac{a_1}{a_2}=\frac{\tan\alpha_2}{\tan\alpha_1} \qquad \text{and} \qquad \cos\alpha_1\cos\alpha_2=0.5$$

In a particular engine m_1 is 600 kg, m_2 is 900 kg, and a_1 is 3 m. Calculate the maximum unbalanced secondary couple if the crank and connecting-rod lengths are 450 mm and 1800 mm respectively and the engine is running at 150 rev/min.
[144 kNm]

8. A four-stroke engine has five identical cylinders in-line and spaced at equal intervals of 150 mm. The reciprocating parts per cylinder are 1.4 kg, the stroke is 100 mm and the connecting-rods are 175 mm between centres. The firing order is 1–4–5–3–2 at equal intervals and the engine speed is 600 rev/min.

Show the primary and secondary forces are balanced and find the maximum primary and secondary couples and the positions of crank 1 when these occur. What is the total unbalanced couple at the instant when crank 1 has turned 10° from top-dead-centre?
[64.7 Nm at 28° and 208°, 56.3 Nm at 25°, 115°, 205° and 295°; 110 Nm]

9. Two alternative designs are being considered for a two-stroke diesel engine having six cylinders in-line and spaced 0.9 m apart. The cranks are to be at 60°, either in order 1–5–3–6–2–4 or 1–4–5–2–3–6. The stroke is to be 250 mm and each connecting-rod 550 mm. The reciprocating parts for each cylinder are 950 kg and the rotating parts 700 kg at the crank radius, and the engine runs at 250 rev/min.

Show that the primary and secondary forces are balanced, and that the secondary moments are balanced with the first arrangement and the primary moments with the second. Calculate the maximum unbalanced moment in each case.
[441 kNm, 115 kNm]

10. A 90° V8 4-stroke engine has cylinders numbered 1, 2, 3, 4 on one bank and 5, 6, 7, 8 on the other, counting from the same end. Show that the firing order 1–5–4–8–6–3–7–2 is consistent with the use of common crank pins and will give uniform firing intervals.

Examine the state of balance.
[Primary and secondary forces, and secondary moments, are all internally balanced. There is a sinusoidally varying primary moment having a maximum value of $\sqrt{(10)}mr\omega^2 d$ on each bank, with a 90° phase difference, so that complete balance could be achieved by means of two opposed masses spaced along the crankshaft to give an equivalent mrd value in a plane 18.4° ahead of crank 1–5]

13 Vibrations of Single-degree-of-freedom Systems

13.1 INTRODUCTION TO VIBRATIONS

A mechanical system is said to be vibrating when its component parts are undergoing periodic (that is cyclically repeated) oscillations about a central configuration (usually the statical equilibrium position). It can be shown that any system, by virtue of its inherent mass and elasticity, can be caused to vibrate by externally applied forces. The duration and severity of the vibration will depend on the relation between the external forces and the mechanics of the system, and will be discussed later.

Although vibration can sometimes be used to advantage—as in cleaning or mixing machines—its presence is generally undesirable for three main reasons. Structural damage of a fatigue nature may be caused by the cyclical fluctuation of loading; physical discomfort may be experienced by personnel associated with the system (for example, passengers in a vehicle); noise, itself a vibration of air molecules, may be generated by a mechanical vibration.

A vibration is characterised and assessed by three parameters, *amplitude*, *frequency* and *phase*. Amplitude is the maximum displacement from the central position (measured as a linear or angular quantity); frequency is the reciprocal of the *period* (the time for one complete cycle of vibration) and is expressed in Hz (cycles/s); phase is a measure of the instant at which a vibration passes through the central position, and in mechanical systems is usually only of importance when the relation between two vibrations, or between the motions of two parts of a system, is being considered.

Free Vibrations

If a mechanical system is displaced from its equilibrium position and then released, the restoring force (arising from either spring elements as in a vehicle suspension, material stiffness as in torsional or bending systems, or gravitational forces as in a pendulum) will cause a return towards the equilibrium position. There will inevitably be an overshoot on the other side, and so on, resulting in what is called a free vibration. This type of vibration arises from an initial input of energy that is continually changing from the potential (or strain energy) form to the kinetic form as the system moves between its extreme positions and its mid-position. In a free vibration the system is said to vibrate at a *natural frequency*.

Damping

Due to various causes there will always be some loss (that is, 'dissipation') of mechanical energy during each cycle of vibration, and this effect is called damping. A free vibration will die away (that is 'decay') due to damping, though if the damping forces are small enough they will have little influence on the frequency and are often neglected to simplify the mathematics.

Degrees of Freedom

The present chapter is restricted to the consideration of single-degree-of-freedom systems, and these can vibrate in only one mode (for example, a pendulum swinging in a vertical plane). In practice many systems have more than one degree-of-freedom (a vehicle may pitch, bounce, or roll) each having its own natural frequency, and these will be analysed in chapter 14. Finally there are the 'continuous' systems, such as beams, where mass and elasticity are inextricably linked and in theory an infinite number of modes of vibration is possible. Some consideration will be given to these cases in chapter 15, though fortunately it is rarely necessary to calculate more than one or two of the lower frequencies at which such systems can vibrate.

Forced Vibrations

When a harmonically varying external force or displacement (known as the 'excitation') is applied to a single-degree-of-freedom system at rest, it is found that the vibration initiated is a combination of one motion at the natural frequency together with one at the 'forcing' frequency. However, the natural frequency component will die out after sufficient time has elapsed for damping to take effect. The system is then said to be performing '*steady-state forced vibrations*'. These occur at the excitation (or forcing) frequency and with an amplitude that depends on the ratio between forcing and natural frequency and the level of damping. When the forcing frequency is equal to the natural frequency the system is said to be in *resonance* and the amplitude may build up to a very high value (limited only by damping and the physical constraints of the system). If the excitation consists of a number of harmonic components at n different frequencies, and the system has N degrees of freedom, then there will be nN possible resonant conditions (that is, values of frequency at which resonance will occur).

Self-excited Vibrations

There are a number of cases where vibrations can be sustained by energy fed in from a *steady* external source (as distinct from the fluctuating excitation required for forced vibrations). The necessary cyclical variations are produced by the motion of the system itself. Examples are stick–slip vibrations (such as brake squeal) arising from the difference between static and kinetic values of friction, also aircraft flutter and other wind-excited vibrations (for example, of bridge structures, power cables and ship's rigging) in which 'lift' forces fluctuate with the attitude and lateral velocity of the vibrating member. The feature of self-excited vibrations is that they take place at a natural frequency of the system, and in this way they can be distinguished from forced vibrations.

From the foregoing it is readily seen that, whatever might be the cause of vibration, the first requirement is to determine the natural frequencies of the system. From this will follow an appreciation of the possible resonant conditions that might arise and a means of diagnosing the source of the excitation. A detailed study of single-degree-of-freedom systems will give an understanding of the principles involved, which can then be extended to the analysis of more complex systems.

13.2   FREE VIBRATIONS WITHOUT DAMPING

Figure 13.1 shows diagrammatically three types of single-degree-of-freedom systems. They are (a) a spring–mass system, (b) a gravity pendulum, and (c) a shaft–inertia system, and they can vibrate in a linear, angular, or torsional mode respectively.

In each of these cases it will be assumed that the vibrating body (or bodies) is rigid, and that any elastic elements are massless. In order to obtain a simple mathematical model it will also be necessary to assume that the restoring force produced by a displacement is proportional to that

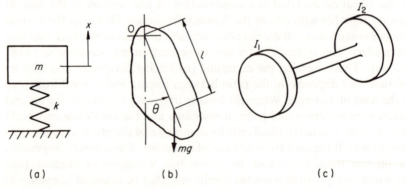

(a)   (b)   (c)

Figure 13.1   Single-degree-of-freedom systems (a) linear ; (b) angular ; (c) torsional

displacement. With these approximations it will then be possible to derive an equation of motion having a basic form that is common to all.

As explained in section 13.1, a free vibration can be initiated by displacing the system from its statical equilibrium position and then releasing it. Newton's second law can be applied directly to the subsequent motion, and this method will now be applied to each of the systems illustrated.

13.2.1 Spring–Mass Vibrations (figure 13.1a)

The mass m is supported on a spring element of stiffness k, assumed constant. Let x be the displacement of m from its equilibrium position; then the net accelerating force on the mass is kx, since the spring force is just sufficient to balance the weight mg when $x = 0$. However, kx is mathematically in the opposite sense to x increasing, so that the equation of motion must be written

$$m\ddot{x} = -kx$$

or

$$\ddot{x} + \omega_n^2 x = 0 \tag{13.1}$$

where

$$\omega_n = \sqrt{(k/m)} \tag{13.2}$$

The solution has two constants of integration and can be written

$$x = A \sin(\omega_n t - \phi) \tag{13.3}$$

where A is the amplitude (and depends entirely on the initial displacement given) and ϕ is a phase angle.

The motion can be displayed graphically as in figure 13.2, where it can be seen that x is given by the projection on to the vertical axis of a vector of magnitude A rotating at angular velocity ω_n. For this reason ω_n is called the *circular frequency* of free vibration. The periodic time of the variation is

$$t_p = 2\pi/\omega_n \tag{13.4}$$

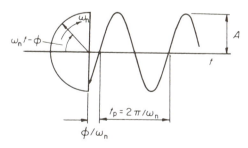

Figure 13.2   *Free undamped vibration*

and hence the natural frequency

$$f_n = \omega_n/2\pi \tag{13.5}$$

$$= \frac{1}{2\pi}\sqrt{\frac{k}{m}} \tag{13.6}$$

(from equation 13.2)

$$= \frac{1}{2\pi}\sqrt{\frac{g}{\delta_s}} \tag{13.7}$$

(where $\delta_s = mg/k$, the static deflection due to gravity).

13.2.2 Pendulum Oscillations (figure 13.1b)

Let m be the mass of the pendulum acting at a distance l from the axis of suspension at O, and I_O be the moment of inertia about O. If the pendulum is given a displacement θ from the vertical, there will be a restoring moment about O equal to $mgl \sin \theta$, and this may be approximated to $mgl\theta$ for small swings. The equation of angular motion about O is then

$$I_O\ddot{\theta} = -mgl\theta$$

or

$$\ddot{\theta} + (mgl/I_O)\theta = 0 \tag{13.8}$$

By comparison with equations 13.1 and 13.5, the natural frequency is

$$f_n = \frac{1}{2\pi}\sqrt{\frac{mgl}{I_O}} \tag{13.9}$$

Example 13.1

A uniform platform 3 m long, 1 m wide and of mass 100 kg carries a small central load of 20 kg and is hinged along a short side as shown in figure 13.3. It is supported in the horizontal position by two springs symmetrically

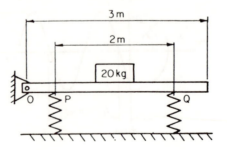

Figure 13.3

placed at P and Q, whose stiffnesses are 20 000 N/m and 5000 N/m respectively.

Find the natural frequency of the platform about its hinge.

If the platform is displaced by an angle θ from its equilibrium position, the additional force in the spring at P is

$$20\,000 \times 0.5\theta = 10\,000\theta \text{ N}$$

and the force in the spring at Q is

$$5000 \times 2.5\theta = 12\,500\theta \text{ N}$$

moment of inertia about $O = 100(3^2/3) + 20 \times 1.5^2$

$$= 345 \text{ kg m}^2$$

The equation of motion about O is then

$$345\,\ddot{\theta} = -(10\,000\theta) \times 0.5 - (12\,500\theta) \times 2.5$$

that is

$$\ddot{\theta} + 105\theta = 0$$

By comparison with equations 13.1 and 13.5

$$f_n = \frac{\sqrt{105}}{2\pi} = 1.63 \text{ Hz}$$

13.2.3   Torsional Vibrations (figure 13.1c)

Figure 13.1c shows a system of two inertias of magnitude I_1 and I_2 supported in 'frictionless' bearings and joined by a 'light' shaft of torsional stiffness k (equal to torque per radian twist, GJ/l for a uniform shaft—see G. H. Ryder, *Strength of Materials*). If θ_1 and θ_2 are the angular displacements of I_1 and I_2 respectively from the untwisted position, then the torque in the shaft is $k(\theta_1 - \theta_2)$ in the sense tending to decrease θ_1 and to increase θ_2 (note that this torque will change sign during a vibration, since θ_1 and θ_2 can take positive and negative values).

In practice torsional vibrations are frequently superimposed on a steady running speed but this does not affect the equations of motion, which are as follows

$$I_1\ddot{\theta}_1 = -k(\theta_1 - \theta_2)$$
$$I_2\ddot{\theta}_2 = k(\theta_1 - \theta_2)$$

Dividing by I_1 and I_2 respectively and subtracting

$$\ddot{\theta}_1 - \ddot{\theta}_2 = -k(\theta_1 - \theta_2)\left(\frac{1}{I_1} + \frac{1}{I_2}\right)$$

Writing $\phi = \theta_1 - \theta_2$, the twist in the shaft

$$\ddot{\phi} + k\left(\frac{I_1 + I_2}{I_1 I_2}\right)\phi = 0 \tag{13.10}$$

giving the natural frequency

$$f_n = \frac{1}{2\pi}\sqrt{\frac{k(I_1 + I_2)}{I_1 I_2}} \tag{13.11}$$

If one end of the shaft is fixed (equivalent to an infinite value of one inertia, say I_2), and the inertia of the free end is I (for I_1) then equation 13.11 becomes

$$f_n = \frac{1}{2\pi}\sqrt{\frac{k}{I}} \tag{13.12}$$

In the two-inertia case, because during vibration the shaft is being wound up first in one direction then in the other, the inertias of the ends must be moving in opposite directions (they are said to be in *anti-phase*, that is, a phase difference of 180°). If their amplitudes are A_1 and A_2, then figure 13.4 shows that there is a point of no twist (called a *node*) in the shaft such that

$$l_1/l_2 = A_1/A_2 \tag{13.13}$$

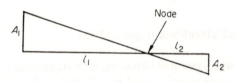

Figure 13.4

The node can be treated as a fixed end, so that if k_1 and k_2 are the stiffnesses of the shaft on either side of the node, equation 13.12 can be written

$$f_n = \frac{1}{2\pi}\sqrt{\frac{k_1}{I_1}} = \frac{1}{2\pi}\sqrt{\frac{k_2}{I_2}}$$

giving

$$k_1/k_2 = I_1/I_2 \tag{13.14}$$

But shaft stiffness is inversely proportional to length, that is

$$k_1/k_2 = l_2/l_1 \tag{13.15}$$

Eliminating k_1/k_2 between equations 13.14, 13.14 and 13.15

$$l_1/l_2 = A_1/A_2 = I_2/I_1 \tag{13.16}$$

13.3   FREE VIBRATIONS WITH DAMPING

Damping forces have been defined as those that will result in a dissipation of energy from a vibrating system. They exist in many forms, for example hysteresis due to non-linear strains within the elastic members, Coulomb damping arising at any rubbing surfaces, air resistance, electromagnetic forces, and viscous fluid forces. Because of the ease of handling the mathematics, it is usual to reduce all damping to an equivalent viscous force (that is, proportional to the velocity at any instant) which will produce the same energy loss per cycle. It can be shown that the vibrational response will be similar and largely independent of the actual sources of damping.

Damping may be inherent in the system (often at a very low level as in structural damping at joints, or in bearing losses) or may be added where essential to the functioning (as the vehicle 'shock absorber', or in instrument damping). Figure 13.5a represents the spring–mass system with damping, where c is the damping coefficient, such that the damping force is c times the velocity. The equation of motion now becomes

$$m\ddot{x} = -kx - c\dot{x} \tag{13.17}$$

which can be written

$$\ddot{x} + 2\zeta\omega_n\dot{x} + \omega_n^2 x = 0 \tag{13.18}$$

where

$$\zeta = c/(2m\omega_n) \tag{13.19}$$

is called the *damping ratio* and lies between 0 and 1 in most practical cases. A free damped vibration then results from the solution of equation 13.18,

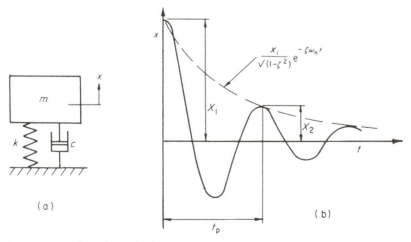

Figure 13.5   *Free damped vibration*

which is

$$x = A e^{-\zeta \omega_n t} \sin(\sqrt{(1 - \zeta^2)} \omega_n t + \phi)$$

(13.20)

The period of damped vibration is

$$t_p = \frac{2\pi}{\sqrt{(1 - \zeta^2)} \omega_n}$$

(13.21)

and the frequency

$$= \sqrt{(1 - \zeta^2)} \omega_n / 2\pi$$

(13.22)

(lower than, but for small damping near to, the natural frequency, equation 13.5). The amplitude decays from cycle to cycle as shown in figure 13.5b, and because the vibration dies away with time it is often called a *transient*.

If the motion is initiated by displacing the mass m a distance X_1 from the equilibrium position and then releasing it, two conditions are obtained from equation 13.20 at $t = 0$

$$X_1 = A \sin \phi$$

$$\dot{x} = 0 = A(-\zeta \omega_n) \sin \phi + A \sqrt{(1 - \zeta^2)} \omega_n \cos \phi$$

giving

$$\tan \phi = \sqrt{(1 - \zeta^2)} / \zeta$$

hence

$$\sin \phi = \sqrt{(1 - \zeta^2)}$$

and

$$A = X_1 / \sqrt{(1 - \zeta^2)}$$

Equation 13.20 then becomes

$$x = \frac{X_1}{\sqrt{(1 - \zeta^2)}} e^{-\zeta \omega_n t} \sin[\sqrt{(1 - \zeta^2)} \omega_n t + \sin^{-1} \sqrt{(1 - \zeta^2)}]$$

After one complete cycle the vibration again reaches a maximum value; that is, at $t = t_p$

$$x = X_2 = \frac{X_1}{\sqrt{(1 - \zeta^2)}} e^{-\zeta \omega_n t_p}$$

giving the ratio of successive amplitudes

$$X_1 / X_2 = e^{\zeta \omega_n t_p}$$

(13.23)

where $\zeta \omega_n t_p$ is called the *logarithmic decrement*.

Equation 13.23 can be extended to any number of complete cycles (or indeed half-cycles to give the numerical amplitude ratio) such that, for n cycles

$$\text{log decrement} = n\zeta\omega_n t_p = n2\pi\zeta/\sqrt{(1-\zeta^2)} \qquad (13.24)$$

This result provides a convenient experimental method for determining the damping ratio in any single-degree-of-freedom system.

Typical values of damping ratio and its effect on frequency and amplitude are given in table 13.1

Table 13.1

Value of ζ	$\dfrac{Damped\ frequency}{Natural\ frequency}$	$\dfrac{X_n}{X_{n+1}}$
0.05 (structural damping)	0.999	1.37
0.35 (vehicle damping)	0.937	10.5
0.67 (instrument damping)	0.742	290

Example 13.2

A flywheel of mass 10 kg and radius of gyration 0.3 m makes rotational oscillations under the control of a torsion spring of stiffness 5 Nm/rad. A viscous damper is fitted and it is found experimentally that the amplitude is reduced by a factor of 100 over any two complete cycles.

Calculate the damping ratio and damping coefficient, and the periodic time of damped oscillation.

Following equations 13.5 and 13.12, the circular frequency

$$\omega_n = \sqrt{\frac{k}{I}} = \sqrt{\frac{5}{10 \times 0.3^2}} = 2.357 \text{ rad/s}$$

From equation 13.24, for 2 cycles

$$\text{log decrement} = 2 \times 2\pi\zeta/\sqrt{(1-\zeta^2)}$$
$$= 12.57\zeta/\sqrt{(1-\zeta^2)}$$

and hence amplitude ratio

$$100 = e^{12.57\zeta/\sqrt{(1-\zeta^2)}}$$

or

$$4.606 = 12.57\zeta/\sqrt{(1-\zeta^2)}$$

giving

$$\zeta = 0.344$$

For a torsional oscillation equation 13.19 becomes

$$c = \zeta \times 2I\omega_n$$

$$= 0.344 \times 2 \times 10 \times 0.3^2 \times 2.357$$

$$= 1.46 \text{ Nm per rad/s}$$

From equation 13.21

$$t_p = \frac{2\pi}{\sqrt{(1 - 0.344^2)} \times 2.357}$$

$$= 2.84 \text{ s}$$

13.4 FORCED VIBRATIONS

The principles of forced vibration can be effectively investigated by analysing the response of a spring–mass system to a simple-harmonic excitation. Figure 13.6 shows such a system acted on by a harmonic disturbance at a frequency of $\omega/2\pi$. In figure 13.6a the disturbance takes the form of a fluctuating force of maximum value F applied directly to the mass (as, for example, an unbalanced engine inertia force); in figure 13.6b a fluctuating displacement of amplitude Y is applied to the free end of the spring (a vehicle riding over an undulating road experiences this type of disturbance).

The equation of motion for figure 13.6a is

$$m\ddot{x} + c\dot{x} + kx = F \cos \omega t \qquad (13.25)$$

hence

$$\ddot{x} + 2\zeta\omega_n\dot{x} + \omega_n^2 x = (F/m) \cos \omega t = (F\omega_n^2/k) \cos \omega t \qquad (13.26)$$

and for figure 13.6b

$$m\ddot{x} + c(\dot{x} - \dot{y}) + k(x - y) = 0$$

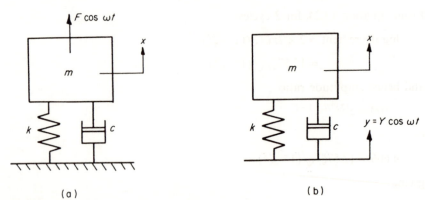

(a)   (b)

Figure 13.6 *Forced vibration of spring–mass system*

or, in terms of the relative displacement $u = x - y$

$$m(\ddot{u} + \ddot{y}) + c\dot{u} + ku = 0$$

which can be written

$$\ddot{u} + 2\zeta\omega_n\dot{u} + \omega_n^2 u = Y\omega^2 \cos \omega t \qquad (13.27)$$

and is considered further later (section 13.4.2).

The general solution of equation 13.26 is made up of a complementary function (the transient vibration, as in section 13.3) together with a particular integral (the steady-state forced vibration), and can be written

$$x = Ae^{-\zeta\omega_n t} \sin(\sqrt{(1 - \zeta^2)}\omega_n t + \phi) + X \cos(\omega t - \beta) \qquad (13.28)$$

where A and ϕ depend on the initial conditions, and X (the forced vibration amplitude) and β (the phase lag) can be determined by substitution of the particular integral into the equation of motion.

Only the steady-state solution will be considered in the following, and substitution of $x = X \cos(\omega t - \beta)$ into equation 13.26 gives

$$-\omega^2 X \cos(\omega t - \beta) - 2\zeta\omega_n \omega X \sin(\omega t - \beta) + \omega_n^2 X \cos(\omega t - \beta)$$

$$= (F\omega_n^2/k) \cos \omega t$$

The terms in this equation can be represented vectorially as in figure 13.7, where the angle $\omega t - \beta$ is measured from an arbitrary datum. It is now a simple step to obtain expressions for X and β from the right-angled triangle containing β.

$$(F\omega_n^2/k)^2 = (\omega_n^2 - \omega^2)^2 X^2 + (2\zeta\omega_n\omega)^2 X^2$$

giving

$$X = \frac{F/k}{\sqrt{[(1 - \omega^2/\omega_n^2)^2 + 4\zeta^2\omega^2/\omega_n^2]}} \qquad (13.29)$$

$$\tan \beta = \frac{2\zeta\omega_n\omega}{\omega_n^2 - \omega^2} = \frac{2\zeta\omega/\omega_n}{1 - \omega^2/\omega_n^2} \qquad (13.30)$$

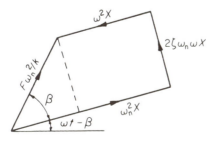

Figure 13.7   Vectorial representation of forced vibration

Equation 13.29 is the amplitude of the forced vibration and can be plotted in non-dimensional form for various values of frequency ratio and damping (figure 13.8a). The ratio

$$\frac{X}{F/k} = \frac{\text{vibrational amplitude}}{\text{static deflection produced by } F}$$

is called the *dynamic amplifier*, and this has a value at resonance (when $\omega/\omega_n = 1$) given by equation 13.29

$$\frac{X}{F/k} = \frac{1}{2\zeta} \tag{13.31}$$

For low levels of damping this can be taken as giving the maximum amplitude of forced vibration, although strictly it can be shown, by differentiation of equation 13.29, that this occurs when

$$\omega/\omega_n = \sqrt{(1 - 2\zeta^2)} \tag{13.32}$$

As the forcing frequency increases above resonance, the amplitude becomes progressively smaller, tending to zero for very high frequencies irrespective of the degree of damping.

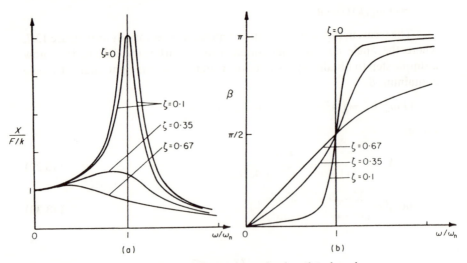

Figure 13.8   Forced vibration response (a) amplitude; (b) phase lag

The variation of phase lag (of the vibration behind the excitation) is given by equation 13.30 and shown graphically in figure 13.8b

It will be seen that

$0 < \beta < 2\pi$ below resonance

$\beta = \pi/2$ at resonance

$\pi/2 < \beta < \pi$ above resonance

Example 13.3

A machine of mass 1000 kg is supported on springs which deflect 8 mm under the static load. With negligible damping the machine vibrates with an amplitude of 5 mm when subjected to a vertical disturbing force of the form $F \cos \omega t$ at 0.8 of the resonant frequency. When a damper is fitted it is found that the resonant amplitude is 2 mm.

Find the magnitude of the disturbing force and the damping coefficient.

$$k = 1000 \times 9.81/0.008 = 1.226 \times 10^6 \text{ N/m}$$

$$\omega_n = \sqrt{(9.81/0.008)} = 35 \text{ rad/s}$$

With zero damping equation 13.29 reduces to

$$X = \frac{F/k}{1 - \omega^2/\omega_n^2}$$

$$0.005 = \frac{F/1.226 \times 10^6}{1 - 0.8^2}$$

giving

$$F = 2207 \text{ N}$$

With damping, and at $\omega = \omega_n$

$$X = \frac{F/k}{2\zeta}$$

$$\zeta = \frac{2207}{1.226 \times 10^6 \times 2 \times 0.002} = 0.45$$

$$= c/2m\omega_n$$

hence

$$c = 0.45 \times 2 \times 1000 \times 35 \text{ N per m/s}$$

$$= 31.5 \text{ kNs/m}$$

13.4.1 Transmissibility and Vibration Isolation

If the vector solution is applied directly to equation 13.25, each vector will represent a maximum force such that the sum of spring force, damping force and inertia force is equal to the excitation. Figure 13.9a shows the configuration for frequencies below resonance, and the diagonal line shown dotted is the vector sum of the spring and damper force, that is, the force transmitted to the foundations which is

$$\sqrt{(k^2 + c^2\omega^2)}X \tag{13.33}$$

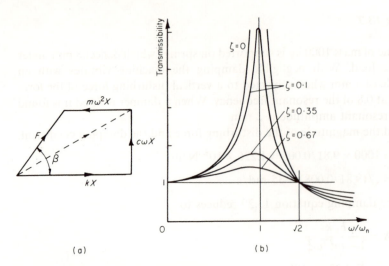

$$(a) \qquad\qquad (b)$$

Figure 13.9   Dynamic force transmission

By substitution from equation 13.29 into 13.33, the *transmissibility*, defined as the ratio

$$\frac{\text{maximum dynamic force transmitted}}{F}$$

$$= \frac{\sqrt{(1 + 4\zeta^2\omega^2/\omega_n^2)}}{\sqrt{[(1 - \omega^2/\omega_n^2)^2 + 4\zeta^2\omega^2/\omega_n^2]}}$$

$$(13.34)$$

This is shown graphically in figure 13.9b, from which it is seen that the transmissibility falls below unity for values of ω/ω_n greater than $\sqrt{2}$. This is the region of vibration isolation, where both the amplitude (figure 13.8a) and the force transmitted are low. It should be noted that, at the higher frequencies, although damping has the effect of reducing the vibrational amplitude it will increase the force transmitted. However, an adequate level of damping is desirable to protect against excessive vibration when passing through resonance (even though the normal operating frequency may be much higher), and to deal with transients.

Example 13.4

A four-cylinder in-line vertical engine with cranks at 180° has a total mass of 200 kg. The reciprocating parts for each cylinder have a mass of 1.5 kg, the connecting-rod is 200 mm between centres and the crank 50 mm.

The engine mounting deflects 4 mm under the static load, and the vibrational amplitude when running at 600 rev/min is 0.1 mm. Determine the dynamic force transmitted to the mounting.

The excitation arises from the unbalanced secondary forces (section 12.5), that is

$$\frac{4mr\omega^2}{n}\cos 2\omega t = \frac{4 \times 1.5 \times 0.05}{4}\left(\frac{600 \times 2\pi}{60}\right)^2 \cos\frac{1200 \times 2\pi}{60}t$$

$$= 296 \cos 125.7t$$

(this represents an excitation of magnitude 296 N at a frequency of 125.7 rad/s)

$$k = 200 \times 9.81/0.004 = 49.05 \times 10^4 \text{ N/m}$$

$$\text{spring force} = kX = 49.05 \times 10^4 \times 0.1 \times 10^{-3}$$

$$= 49.05 \text{ N}$$

$$\text{damping force} = c \times 125.7 \times 0.1 \times 10^{-3}$$

$$= 0.01257c \text{ N}$$

$$\text{inertia force} = 200 \times 125.7^2 \times 0.1 \times 10^{-3}$$

$$= 316 \text{ N}$$

Applying these values to figure 13.9a and solving

$$296^2 = (316 - 49.05)^2 + (0.01257c)^2$$

giving

$$c = 10\,160 \text{ N s/m}$$

The maximum dynamic force transmitted to the foundations is then

$$\sqrt{[(\text{spring force})^2 + (\text{damping force})^2]}$$

$$= \sqrt{[49.05^2 + (0.01257 \times 10\,170)^2]}$$

$$= 137 \text{ N}$$

The reader should note that in this example the excitation is above resonance and that $\beta > 90°$ in the vector diagram. Also the *total* force transmitted to the foundations is the sum of the static and dynamic forces, and will vary between 2099 N and 1825 N.

13.4.2 Vibration-measuring Instruments

The steady-state solution of equation 13.27 derived from figure 13.6b will be of the form

$$u = U \cos(\omega t - \beta)$$

where, by analogy with equation 13.29

$$U = \frac{Y(\omega^2/\omega_n^2)}{\sqrt{[(1 - \omega^2/\omega_n^2)^2 + 4\zeta^2\omega^2/\omega_n^2]}} \tag{13.35}$$

Figure 13.10 shows the variation of U/Y against frequency for zero damping and for 0.67 damping ratio. It can be seen that, for $\omega/\omega_n > 1.5$

$$U \approx Y$$

(the closest approximation and the widest frequency range being obtained for $\zeta = 0.67$).

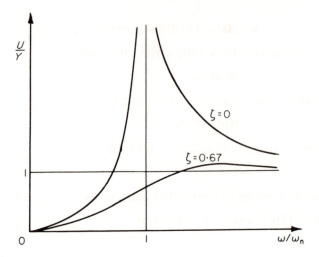

Figure 13.10

This result is made use of in a *seismic-type vibrometer*, in which the seismic mass is spring-mounted inside a rigid case. When the case is attached to the vibration being measured, the relative movement u of the seismic mass is recorded (either by mechanical or electrical means) and hence the frequency and amplitude of the main vibration y is determined.

For $\omega/\omega_n < 0.7$ and $\zeta = 0.67$ equation 13.35 reduces to

$$U \approx \text{constant} \times Y$$

so that an instrument which records U measures a quantity proportional to the maximum acceleration of the vibration $Y \cos \omega t$. This type of instrument (measuring vibrations with frequencies below its own natural frequency) is called an *accelerometer*.

SUMMARY

Free undamped vibrations

Form of general solution $x = A \sin(\omega_n t - \phi)$

Natural frequency $\omega_n = 2\pi f_n$

Periodic time $t_p = 2\pi/\omega_n$

Free damped vibration

Form of general solution

$$x = A e^{-\zeta \omega_n t} \sin(\sqrt{(1 - \zeta^2)}\omega_n t + \phi)$$

Frequency of vibration $= \sqrt{(1 - \zeta^2)}\omega_n/2\pi$

Periodic time $t_p = 2\pi/\sqrt{(1 - \zeta^2)}\omega_n$

Logarithmic decrement $\log_e\left(\dfrac{X_1}{X_2}\right) = \dfrac{2\pi\zeta}{\sqrt{(1 - \zeta^2)}}$

Forced vibration with harmonic force disturbance $F \cos \omega t$

Form of steady-state solution

$$x = X \cos(\omega t - \beta)$$

$$X = \frac{F/k}{\sqrt{[(1 - \omega^2/\omega_n^2)^2 + 4\zeta^2\omega^2/\omega_n^2]}}$$

$$\tan \beta = \frac{2\zeta\omega/\omega_n}{1 - \omega^2/\omega_n^2}$$

Dynamic amplifier $= kX/F$

$$\text{Transmissibility} = \sqrt{\left[\frac{1 + 4\zeta^2\omega^2/\omega_n^2}{(1 - \omega^2/\omega_n^2)^2 + 4\zeta^2\omega^2/\omega_n^2}\right]}$$

Forced vibration with harmonic support disturbance $Y \cos \omega t$

Form of steady-state solution

$$u = U \cos(\omega t - \beta) \qquad \text{where } u = x - y$$

$$U = \frac{Y\omega^2/\omega_n^2}{\sqrt{[(1 - \omega^2/\omega_n^2)^2 + 4\zeta^2\omega^2/\omega_n^2]}}$$

13.5   PROBLEMS

1. A connecting-rod of mass 1 kg is 0.3 m between centres, the big-end bearing being 80 mm diameter and the small-end bearing 20 mm diameter. When suspended on a horizontal knife-edge and oscillated as a pendulum the period is 0.934 s about the big-end and 1 s about the small-end.

   Determine the position of the centre of gravity and the moment of inertia about it.

   [0.19 m from small-end centre, 0.01 kg m^2]

2. A uniform shaft 50 mm diameter is 3 m long and is held rigidly at both ends with its axis vertical. A concentric disc of mass 500 kg and radius of gyration 0.5 m is rigidly attached at a point 2 m from the lower end. Determine the natural frequency of torsional oscillations if the modulus of rigidity is 84 000 N/mm^2.

   [3.96 Hz]

3. Two rotors, one of mass 20 kg and radius of gyration 120 mm, the other 30 kg and 150 mm, are mounted at opposite ends of a shaft of 50 mm diameter. If the frequency of torsional vibration is 100 Hz what is the length of shaft and the position of the node? Modulus of rigidity = 80 000 N/mm^2.

   [616 mm; node at 194 mm from larger rotor]

4. A load of 50 kg is suspended from a spring of stiffness 5 kN/m, and its motion is damped by a force proportional to the velocity, of magnitude 75 N at 1 m/s. If the load is displaced from its equilibrium position and released, find the number of complete vibrations before the amplitude is less than one-tenth of the original displacement.

   [5]

5. A vertical spring of stiffness 10 kN/m supports a mass of 40 kg. There is a constant friction force of 100 N opposing motion, whether upwards or downwards. If the mass is released from a position in which the total extension of the spring is 120 mm determine (a) the time before the mass finally comes to rest, and (b) the final extension of the spring.

   [0.795 s; 40 mm]

6. A mass of 50 kg is supported on springs which deflect statically by 20 mm. A viscous damper with a coefficient of 800 N s/m is fitted. Calculate the frequency of damped vibration and the damping ratio.

   If the mass is then subjected to a periodic force having a maximum value of 200 N with a frequency of 2 Hz, find the amplitude of forced vibration.

   [3.28 Hz, 0.362; 10.3 mm]

7. A spring–mass system has a natural frequency of 1.2 Hz and damping may be neglected. When the mass is stationary the free end of the spring is given a displacement $y = 50 \sin 2\pi t$ mm, where t s is the time from the beginning of the motion. Determine the displacement of the mass after the first 0.3 seconds.

   [53 mm]

8. A single-cylinder engine has a mass of 100 kg and a vertical unbalanced force of $400 \sin 13\pi t$ N. It is supported by springs of total stiffness 60 kN/m and a viscous damper is fitted which gives a force of 700 N per m/s. Determine the maximum force transmitted to the ground through the spring and damper.

   [1220 N]

9. A vehicle is towing a two-wheeled trailer at a speed of 60 km/hr over a road whose surface is approximately sinusoidal, of wavelength 15 m and double-amplitude 50 mm. The mass of the trailer is 400 kg and it is supported by springs of total stiffness 20 kN/m and fitted with dampers giving a damping ratio of 0.4. Find the magnitude of the dynamic force in the springs and the amplitude of the trailer vibration.

   [±616 N; 31.6 mm]

10. A spring–mass accelerometer has a natural frequency of 10 Hz and is damped at 0.6 times the critical value. When attached to a body vibrating at 6 Hz the (relative) amplitude recorded by the instrument is 1 mm. Calculate the amplitude and maximum acceleration of the body.

    [2.67 mm, 3.79 m/s^2]

14 Vibrations of Multi-degree-of-freedom Systems

14.1 INTRODUCTION

The number of *degrees of freedom* of a vibrating system can be defined as the number of independent co-ordinates required to specify its configuration at any instant. As a natural development from the single-degree-of-freedom systems analysed in the previous chapter, particular attention will be paid to the treatment of two-degree-of-freedom vibrations. The methods developed can then be extended to the multi-degree-of-freedom cases.

There are many real systems which, subject to simplifying assumptions, can be approximated with reasonable accuracy to a mathematical model possessing only two degrees of freedom. Three important examples are shown diagrammatically in figure 14.1. For instance, the linear two-mass–spring system of figure 14.1a can be used to represent a vehicle trailer mass, its suspension, the axle mass and the tyre stiffness. Figure 14.1b provides a model for a vehicle mounted on front and rear springs, and the pitch and bounce modes can be analysed by treating the body as rigid and neglecting the 'unsprung' mass. Many drive-transmission systems can be reduced to the three-inertia torsional case of figure 14.1c, using modified values of inertia

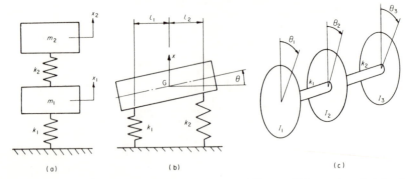

Figure 14.1   Two-degree-of-freedom system (a) linear; (b) angular; (c) torsional

and shaft stiffness where a gear ratio is included in the system (note that the displacements shown can be reduced to two independent co-ordinates in terms of the twist in each length of shaft).

It can be shown mathematically that the number of natural frequencies of a system is equal to the number of degrees of freedom, each being associated with a *principal* or *normal mode* of free vibration. In a principal mode all components of the system vibrate with the same frequency and reach their extreme positions at the same instant. It then follows that any arbitrary free vibration must consist of a combination of separate vibrations (in the correct proportions and phasing) in the principal modes. When an external excitation is applied to the system a forced vibration will be set up and resonance will occur if the excitation frequency coincides with any natural frequency. Consequently the first and most important step in determining the vibration response is to estimate the natural frequencies of the system, and for this purpose the effects of damping need not be considered.

14.2   FREE VIBRATIONS OF TWO-DEGREE-OF-FREEDOM SYSTEMS

Before dealing with a particular example it will be useful to outline the procedure to be followed in analysing the appropriate mathematical model.

(a) Consider the system in a configuration displaced from equilibrium, and choose the most suitable parameters (that is, co-ordinates) to specify this position and to facilitate the formulation of the equations of motion.
(b) Write down the equations of motion of each mass or inertia as appropriate, in terms of the chosen co-ordinates.
(c) Assume solutions to the equations of motion that will represent the principal modes. Because no damping term has been included in the equations of motion, all parts of the system will vibrate in-phase or anti-phase, and simple harmonic solutions of the form $A_1 \sin \omega_n t$, $A_2 \sin \omega_n t$, ... (equal to the number of degrees-of-freedom) will suffice.
(d) Substitute the assumed solutions into the equations of motion and eliminate the amplitude ratios $A_1 : A_2 : \ldots$ between them.
(e) A frequency equation now results, and the roots of this equation are the values of ω_n in the principal modes.
(f) The modal shapes can now be found by substituting each value of ω_n into the expressions for amplitude ratio at (d).

Applying these steps to the system of figure 14.1a, where the displacements x_1 and x_2 are measured from the static equilibrium position, the equations of motion are

$$m_1 \ddot{x}_1 = -k_1 x_1 + k_2 (x_2 - x_1) \tag{14.1}$$

$$m_2 \ddot{x}_2 = -k_2 (x_2 - x_1) \tag{14.2}$$

Making the substitutions $x_1 = X_1 \sin \omega_n t$ and $x_2 = X_2 \sin \omega_n t$

$$-m_1 X_1 \omega_n^2 = -k_1 X_1 + k_2(X_2 - X_1)$$

$$-m_2 X_2 \omega_n^2 = -k_2(X_2 - X_1)$$

Re-arranging each equation to give the amplitude ratio

$$\frac{X_1}{X_2} = \frac{k_2}{k_1 + k_2 - m_1 \omega_n^2} \tag{14.3}$$

$$\frac{X_1}{X_2} = \frac{k_2 - m_2 \omega_n^2}{k_2} \tag{14.4}$$

Eliminating X_1/X_2 between equations 14.3 and 14.4

$$k_2^2 = (k_1 + k_2 - m_1 \omega_n^2)(k_2 - m_2 \omega_n^2)$$

giving the frequency equation

$$m_1 m_2 \omega_n^4 - [(m_1 + m_2)k_2 + m_2 k_1]\omega_n^2 + k_1 k_2 = 0 \tag{14.5}$$

Example 14.1

If in the system shown in figure 14.1a, $m_1 = 5$ kg, $m_2 = 2$ kg, $k_1 = 5$ kN/m and $k_2 = 10$ kN/m, calculate the natural frequencies and the amplitude ratio in the principal modes.

From equation 14.5

$$5 \times 2\omega_n^2 - (7 \times 10\,000 + 2 \times 5000)\omega_n^2 + 5000 \times 10\,000 = 0$$

$$\omega_n^4 - 8000\omega_n^2 + 5 \times 10^6 = 0$$

$$\omega_n^2 = 683 \text{ or } 7317$$

$$\omega_n = 26.1 \text{ or } 85.5 \text{ rad/s}$$

$$f_n = 4.16 \text{ or } 13.6 \text{ Hz}$$

The amplitude ratio in the two modes can be obtained by substituting the two values of ω_n^2 into equation 14.4 (equation 14.3 would give the same numerical value)

$$\frac{X_1}{X_2} = \frac{10\,000 - 2 \times 683}{10\,000} = 0.863$$

(the first mode, in-phase)

$$\frac{X_1}{X_2} = \frac{10\,000 - 2 \times 7317}{10\,000} = -0.463$$

(the second mode, anti-phase).

14.3   FORCED VIBRATIONS OF TWO-DEGREE-OF-FREEDOM SYSTEMS

If a harmonic excitation is applied to a two-degree-of-freedom system a steady-state forced vibration will result (after the transient motion has died away) in much the same manner as in the single-degree-of-freedom case. All parts of the system will then be vibrating at the forcing frequency and with amplitudes proportional to the magnitude of the excitation. The response at any particular frequency will depend on the value of the forcing frequency in relation to the natural frequencies, resonance occurring when it coincides with either value.

For purposes of illustration the two-mass–spring system will be considered, and damping will be neglected to simplify the mathematics. There are several ways in which the excitation can be applied (either as a fluctuating force or displacement), but an important case arises when m_1 in figure 14.1a is acted on by a harmonic force which may be written $F \cos \omega t$. The equations of motion are now

$$m_1 \ddot{x}_1 = -k_1 x_1 + k_2(x_2 - x_1) + F \cos \omega t \tag{14.6}$$

$$m_2 \ddot{x}_2 = -k_2(x_2 - x_1) \tag{14.7}$$

The steady-state vibration is given by the particular integral solution, of the form

$$x_1 = X_1 \cos \omega t$$

$$x_2 = X_2 \cos \omega t$$

Substituting in equations 14.6 and 14.7 and solving

$$X_1 = \frac{(k_2 - m_2 \omega^2)F}{m_1 m_2 \omega^4 - [(m_1 + m_2)k_2 + m_2 k_1]\omega^2 + k_1 k_2} \tag{14.8}$$

$$X_2 = \frac{k_2 F}{m_1 m_2 \omega^4 - [(m_1 + m_2)k_2 + m_2 k_1]\omega^2 + k_1 k_2} \tag{14.9}$$

It will be seen that the denominator in these expressions is the same as the left-hand side of the frequency equation 14.5, showing that the amplitudes are infinite when $\omega = \omega_n$. The response over the complete frequency range is shown in figure 14.2. The important feature of this response is that when $\omega = \sqrt{(k_2/m_2)}$ the amplitude of vibration of m_1 is zero. This property can be utilised to suppress the vibration of m_1 at a particular forcing frequency, and is known as the principle of the vibration absorber to be discussed in the next section.

14.4   VIBRATION ABSORBER

It can be deduced from the results of section 14.3 (and in particular

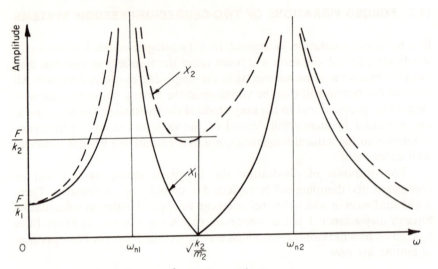

Figure 14.2 Forced vibration of two-mass–spring system

equation 14.8) that if a single-degree-of-freedom system (for example m_1, k_1) is subjected to a harmonic excitation, the resulting vibration can be nullified by attaching to the main system a secondary system (m_2, k_2) such that the natural frequency of the second system on its own $(\sqrt{(k_2/m_2)})$ is equal to the forcing frequency. This second system is called a dynamic vibration absorber.

The physical explanation is that, in the steady state, the motion is such that the force in the spring k_2 is always equal and opposite to the excitation, so m_1 remains stationary. However, this simple type of absorber is only effective at one frequency (usually chosen to be the resonant frequency of the main system), and by introducing a second degree of freedom two new resonant frequencies arise. If damping is added to the absorber system a compromise solution is obtained which restricts the amplitude of the main mass over a wide frequency range (including the resonant frequencies) but no longer passes through a zero value.

14.5 TORSIONAL VIBRATIONS

There are two aspects in which torsional vibrations differ from those of most mass–spring systems, and consequently merit separate treatment. In the first place they are usually superimposed on a steady running speed, though this may be disregarded in formulating the equations of motion (displacements being reckoned from a quasi-static unstrained position). Secondly the system is isolated from the 'fixed' earth through its bearings, so that all parts of the system are free to move.

It therefore appears at first sight that the three-inertia system of figure 14.1c will have three degrees of freedom (and it is, in fact, convenient to specify the displacements of each inertia as the variable co-ordinates). However, it can be shown that there are only two independent variables, these being the twists in the connecting shafts (that is, the differences between the inertia displacements at the end of each shaft).

14.5.1   Two-degree-of-freedom Torsional Vibrations

Referring to figure 14.1c, the equations of motion can be written

$$I_1 \ddot{\theta}_1 = k_1 (\theta_2 - \theta_1) \tag{14.10}$$

$$I_2 \ddot{\theta}_2 = -k_1 (\theta_2 - \theta_1) + k_2 (\theta_3 - \theta_2) \tag{14.11}$$

$$I_3 \ddot{\theta}_3 = -k_2 (\theta_3 - \theta_2) \tag{14.12}$$

It is possible to rearrange these equations in terms of $\theta_2 - \theta_1$, $\dot{\theta}_2 - \dot{\theta}_1$, etc., and hence reduce to two independent equations in ϕ_1 and ϕ_2 where these are the twists in the shafts. However, it is not essential to do this and the natural frequencies can be obtained by following the procedure of section 14.2. Assume

$$\theta_1 = A_1 \sin \omega_n t$$

$$\theta_2 = A_2 \sin \omega_n t$$

$$\theta_3 = A_3 \sin \omega_n t$$

Substituting in the equations of motion gives

$$-I_1 \omega_n^2 A_1 = k_1 (A_2 - A_1) \tag{14.13}$$

$$-I_2 \omega_n^2 A_2 = -k_1 (A_2 - A_1) + k_2 (A_3 - A_2) \tag{14.14}$$

$$-I_3 \omega_n^2 A_3 = -k_2 (A_3 - A_2) \tag{14.15}$$

From equation 14.13

$$A_1 = \frac{k_1 A_2}{k_1 - I_1 \omega_n^2} \tag{14.16}$$

From equation 14.15

$$A_3 = \frac{k_2 A_2}{k_2 - I_3 \omega_n^2} \tag{14.17}$$

Substituting for A_1 and A_3 from equations 14.16 and 14.17 into equation 14.14

$$-I_2 \omega_n^2 A_2 = -k_1 A_2 \left(1 - \frac{k_1}{k_1 - I_1 \omega_n^2} \right) + k_2 A_2 \left(\frac{k_2}{k_2 - I_3 \omega_n^2} - 1 \right)$$

which results in the frequency equation

$$\omega_n^4 - \left[k_1 \left(\frac{1}{I_1} + \frac{1}{I_2} \right) + k_2 \left(\frac{1}{I_2} + \frac{1}{I_3} \right) \right] \omega_n^2 + \frac{k_1 k_2}{I_1 I_2 I_3} (I_1 + I_2 + I_3) = 0$$

(14.18)

Example 14.2

In the system shown in figure 14.1c $I_1 = 100$ kg m^2, $I_2 = 150$ kg m^2 and $I_3 = 1000$ kg m^2. The shaft between I_1 and I_2 is 1 m long and 100 mm diameter, and between I_2 and I_3 is 0.6 m long and 50 mm diameter. The modulus of rigidity is 82 000 N/mm^2.

Determine the natural frequencies of torsional vibrations and the positions of the node (or nodes) in each case.

The stiffnesses are

$$k_1 = \frac{82\,000 \times \pi \times 100^4}{1000 \times 32 \times 10^3} = 805\,000 \text{ Nm/rad}$$

$$k_2 = \frac{82\,000 \times \pi \times 50^4}{600 \times 32 \times 10^3} = 83\,860 \text{ Nm/rad}$$

Substituting in equation 14.18

$$\omega_n^4 - \left[805\,000 \left(\frac{1}{100} + \frac{1}{150} \right) + 83\,860 \left(\frac{1}{150} + \frac{1}{1000} \right) \right] \omega_n^2$$

$$+ \frac{805\,000 \times 83\,860 \times 1250}{100 \times 150 \times 1000} = 0$$

$$\omega_n^4 - 14\,060 \omega_n^2 + 5.626 \times 10^6 = 0$$

$$\omega_n^2 = 412 \text{ or } 13\,650$$

$$\omega_n = 20.3 \text{ or } 117 \text{ rad/s}$$

$$f_n = 3.23 \text{ or } 18.6 \text{ Hz}$$

For the first mode $\omega_n = 20.3$ rad/s and there will be one node, whose position can be found from the amplitude ratios (and lies between those inertias that are vibrating in anti-phase). It follows from equation 14.17 that the node is between I_2 and I_3, since

$$\frac{A_3}{A_2} = \frac{83\,860}{83\,860 - 1000 \times 412} = -0.256$$

By proportion, the distance of the node from I_3 is then

$$\frac{0.256}{1.256} \times 600 = 122 \text{ mm}$$

For the second mode $\omega_n = 117$ rad/s and there will be a node between each inertia. Equations 14.16 and 14.17

$$\frac{A_1}{A_2} = \frac{805\,000}{805\,000 - 100 \times 13\,650} = -1.4375$$

$$\frac{A_3}{A_2} = \frac{83\,860}{83\,860 - 1000 \times 13\,650} = -0.0062$$

From these ratios, the nodes occur at 590 mm from I_1 and 3.7 mm from I_3.

14.5.2   Vibrations of Geared Torsional Systems

In order to calculate the natural frequency of a geared system such as figure 14.3a it is necessary first of all to reduce it to an equivalent in-line system.

If G is the gear ratio between the speed of shaft 1 and shaft 2, then figure 14.3b shows the values of the equivalent system as if acting in line with shaft 2. It will be seen that, where inertia or stiffness has been 'transferred', each item has been multiplied by G^2. This can be justified on the grounds of equal vibrational energy in the two systems, since kinetic energy is of the form $\frac{1}{2}I\omega^2$ and strain energy is of the form $\frac{1}{2}k\theta^2$, both ω and θ being in the ratio G on opposite sides of the gear reduction.

14.6   MULTI-DEGREE-OF-FREEDOM VIBRATIONS

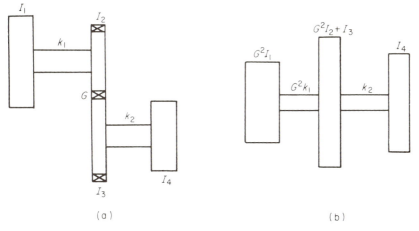

(a)   (b)

Figure 14.3   Geared torsional system

Although it is possible to extend the analytical method of section 14.2 to any number of degrees of freedom, the complexity of the computation increases considerably. In practice it is usually only the lower natural frequencies (often only the first or 'fundamental' mode) that are required, the higher modes being well above the range of excitation frequencies. Consequently a more direct method, attributed to Holzer, which lends itself to numerical computation, is to be preferred. In this method an assumed value of natural frequency is taken as a trial and the mechanics of the system are checked against the boundary conditions. The exact value of natural frequency is then found by successive iteration. As an illustration the six-inertia torsional system of figure 14.4 will be considered. Let A_1, A_2, ..., A_6 be the amplitudes of $I_1, I_2 \cdots I_6$ in a principal mode having a circular frequency ω_n. Then $\theta_1 = A_1 \sin \omega_n t$, $\theta_2 = A_2 \sin \omega_n t$ ..., and following equation 14.13

$$A_2 = A_1 - I_1 A_1 \omega_n^2 / k_1$$

equations 14.13 and 14.14 together give

$$A_3 = A_2 - (I_1 A_1 + I_2 A_2)\omega_n^2 / k_2$$

Extending this process to the end of the six-inertia system

$$A_6 = A_5 - (I_1 A_1 + I_2 A_2 + \ldots I_5 A_5)\omega_n^2 / k_5 \qquad (14.19)$$

Now, if I_6 is a 'free' end

$$-I_6 A_6 \omega_n^2 = -k_5(A_6 - A_5)$$

and, by substitution in equation 14.19

$$\sum_1^6 IA = 0 \qquad (14.20)$$

(the condition for a 'fixed' end would be $A_6 = 0$).

In any numerical problem the application of the appropriate end condition serves as a check on whether the assumed value of frequency coincides with a principal mode.

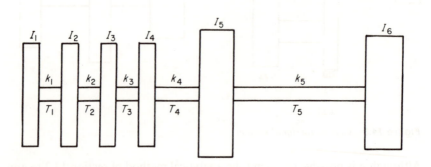

Figure 14.4 Multi-rotor torsional system

Example 14.3

Figure 14.4 represents a four-cylinder engine and flywheel driving a load, and the effective values of inertia and torsional stiffness are $I_1 = I_2 = I_3 = I_4 = 60$ kg m^2, $I_5 = 700$ kg m^2, $I_6 = 300$ kg m^2, $k_1 = k_2 = k_3 = 15 \times 10^6$ Nm/rad, $k_4 = 16 \times 10^6$ Nm/rad, $k_5 = 2.5 \times 10^6$ Nm/rad. Calculate the fundamental frequency of torsional vibrations.

Since only the first mode frequency is required, an estimate may be obtained by reducing the system to two inertias, say 800 kg m^2 and 300 kg m^2 joined by a torsional stiffness of 2×10^6 Nm/rad. From equation 13.11, this would have a natural frequency

$$\omega_n = \sqrt{\frac{2 \times 10^6 \times 1100}{800 \times 300}} = 96 \text{ rad/s}$$

The procedure of the Holzer method will now be followed, taking $\omega_n = 100$ rad/s for a first trial, and letting $A_1 = 1$ unit for simplicity. It is convenient to enter the values and calculations in a table as each step is carried out.

I	k	ω_n^2/k	A	IA	ΣIA	$\delta A = -\Sigma IA\omega_n^2/k$
60			1	60	60	
	15×10^6	6.67×10^{-4}				-0.040
60			0.96	57.6	117.6	
	15×10^6	6.67×10^{-4}				-0.0784
60			0.88	52.9	170	
	15×10^6	6.67×10^{-4}				-0.1137
60			0.768	46.1	216	
	16×10^6	6.25×10^{-4}				-0.135
700			0.633	443	659	
	2.5×10^6	40×10^{-4}				-2.636
300			-2.0	-600	59	

Since there is only one node and A_1 has been assumed positive, the 'remainder' 59 for ΣIA (which should be zero) indicates that the assumed value for ω_n is too low. The calculations can be repeated for higher values and it will be found that the actual frequency is

$\omega_n = 103.4$ rad/s

$f_n = 16.5$ Hz

SUMMARY

Degrees of freedom

Free vibrations of two degree of freedom systems

   Principal mode solutions

   Amplitude ratios

   Frequency equation and natural frequencies

Forced vibration of two degree of freedom systems

   Steady-state response

   Vibration absorber

Torsional systems

Multi-degree of freedom systems

   Holzer method

14.7   PROBLEMS

1. A two-wheeled trailer has a total mass of 600 kg, of which the wheel and axle assembly is 100 kg. The suspension springs have a total stiffness of 90 N/mm and each tyre has a stiffness of 120 N/mm.

   Determine the forward speeds at which resonance will occur if towed along a road having an undulating surface of wavelength 4 m.

   [7.23 m/s, 36.85 m/s]

2. A motor having a mass of 60 kg is supported elastically on a mounting which deflects 5 mm under the static load. Due to rotating unbalance at the motor running-speed of 400 rev/min a vibration is set up having an amplitude of 4 mm. This is to be eliminated by attaching to the motor a spring carrying at its free end a mass of 2 kg. Damping may be neglected.

   Determine the spring stiffness and the amplitude of vibration of the added mass.

   [3510 N/m, 14.2 mm]

3. If, in the system shown in figure 14.1b, $l_2 = 2l_1$, the mass is m and the radius of gyration about G is l_1, and each spring is of stiffness k, determine the natural frequencies in the two principal modes.

   [$0.207 \sqrt{k/m}$, $0.367 \sqrt{k/m}$ Hz]

4. If a vertical excitation of the form $F \cos \omega t$ is applied to G in problem 3, show that when $\omega^2 = 5\, k/m$ the amplitude of x is zero, and find the amplitude of θ.

   [F/kl_1]

5. An inertia I_1 of value 150 kg m², is connected by a shaft 0.76 m long and 75 mm diameter to a second inertia I_2 of value 200 kg m². I_2 is connected by a shaft 0.5 m long and 50 mm diameter to a fixed framework. Find the frequencies of torsional vibration of the system and the twist in each shaft if the amplitude of I_1 is 1°. The modulus of rigidity is 82 000 N/mm².

   [2.62, 10.24 Hz; 0.12°, 0.88°; 1.85°, 0.85°]

6. A turbine of moment of inertia 4 kg m² is connected by a shaft of stiffness 10 kNm/rad to a pinion of inertia 0.25 kg m². The pinion drives a wheel of inertia 160 kg m² with a speed reduction of 8:1, and this is connected by a shaft of stiffness 1.3 MNm/rad to a generator of inertia 60 kg m². Find the natural frequencies of torsional vibration of the generator shaft and the amplitude ratio between the turbine and generator in each principal mode.

   [11.3, 27.6 Hz; −6.10, 0.284]

7. A torsional system consists of three inertias I_1, I_2 and I_3 in line and of magnitudes 10, 20 and 30 kg m² respectively. The torsional stiffness of the coupling between I_1 and I_2 is 20 kNm/rad, between I_2 and I_3 is 50 kNm/rad and I_3 is connected through a stiffness of 100 kNm/rad to a fixed frame. Show by the Holzer method that the fundamental frequency is approximately 4.5 Hz.

8. For the system of example 14.3 in section 14.6, show that the second mode frequency of torsional vibrations is approximately 32 Hz.

15 Lateral Vibrations and Whirling Speeds

15.1 INTRODUCTION

In the types of vibration considered in previous chapters it has always been possible to assume, without introducing undue errors, that the system consisted of one or more rigid bodies connected by massless elastic elements. Following from this it became relatively straightforward to formulate the equations of motion in terms of the chosen displacement co-ordinates, and to solve for the natural frequencies according to the number of degrees of freedom. For various reasons this approach cannot be applied directly to the vibrations of beams and shafts. Both mass and elasticity are generally 'distributed' along the length and mathematically there are then infinite degrees of freedom. Even when 'point' masses predominate and the mass of the beam is neglected it is not possible (except in the single-mass case) to write down the restoring force on any mass in terms of its displacement at any instant. In consequence, 'approximate' methods have been developed for general use, 'exact' analysis being applied only to a few standard cases. These approximate methods are usually limited to the determination of the fundamental (that is, first mode) frequency which, as in most multi-degree-of-freedom vibrations, is of prime practical importance.

It is instructive to examine the exact methods first, since the solutions give an insight into the modes of vibration and will indicate suitable assumptions (for example, of vibration configuration) that can be applied when using the approximate methods. It will also enable an assessment to be made of the errors involved in alternative calculations of natural frequencies.

The phenomenon of *whirling* of shafts arises from the tendency of centrifugal effects due to rotation to deflect the shaft from its bearing centre-line, and at certain 'critical' speeds a form of resonance occurs.

As will become apparent later, the mechanics of whirling is mathematically identical to that of beam vibrations (if gyroscopic effects are neglected) although it becomes more complex if the shaft or bearing stiffnesses are not symmetrical.

15.2 LATERAL VIBRATIONS OF BEAMS

It is much too difficult to attempt an analysis of the general case of a beam, allowing for both distributed and point masses, so these two will be dealt with separately. First an exact solution for point masses, neglecting the beam mass, will be developed, and then the mathematical solution for uniformly distributed mass will be outlined. Following this, some of the alternative approximate methods will be considered.

15.2.1 Single Mass on Light Beam

This is a single-degree-of-freedom system, and if the restoring force on the mass due to beam stiffness is ky, where y is the vibrational displacement, then

$$m\ddot{y} + ky = 0$$

$$\omega_n = \sqrt{(k/m)}$$

$$= \sqrt{(g/\delta)}$$

where δ is the static deflection which would result from an applied load equal to the weight mg (for example, $\delta = mga^2b^2/3EIl$ for a simply supported beam of length l in which m is at distances a and b from the supports; $\delta = mgl^3/3EI$ for a cantilever, where l is the distance of m from the fixed end).

15.2.2 Several Masses on Light Beam—Exact Solution

If positions along the beam are denoted by the x-co-ordinate, then the deflection at a point x due to a unit load applied at x_q can be written a_{pq}. This is defined as an *influence coefficient*, and its value can be calculated from beam-deflection theory when the boundary conditions are given (that is, the positions and types of support).

By applying the principle of superposition, influence coefficients provide a convenient means of expressing the deflection at any point due to combined loading. The application of this to vibrations will be illustrated by an example.

Example 15.1

Determine the lowest natural frequency of the beam illustrated in figure 15.1 which carries mass m, $4m$ and $2m$ at the quarter points and is simply supported at each end.

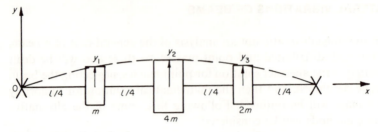

Figure 15.1   Point loads

The vibration of the three masses from the static equilibrium position may be written

$$y_1 = Y_1 \sin \omega_n t \qquad y_2 = Y_2 \sin \omega_n t \qquad y_3 = Y_3 \sin \omega_n t$$

and the maximum inertia forces are then

$$m\omega_n^2 Y_1 \qquad 4m\omega_n^2 Y_2 \qquad 2m\omega_n^2 Y_3$$

Now, if these inertia forces are treated as 'loads' deflecting the beam into its vibrating form, the use of influence coefficients gives

$$Y_1 = a_{11} m\omega_n^2 Y_1 + a_{12} 4m\omega_n^2 Y_2 + a_{13} 2m\omega_n^2 Y_3 \tag{15.1}$$

$$Y_2 = a_{21} m\omega_n^2 Y_1 + a_{22} 4m\omega_n^2 Y_2 + a_{23} 2m\omega_n^2 Y_3 \tag{15.2}$$

$$Y_3 = a_{31} m\omega_n^2 Y_1 + a_{32} 4m\omega_n^2 Y_2 + a_{33} 2m\omega_n^2 Y_3 \tag{15.3}$$

where it can be shown that in this case

$$a_{11} = a_{33} = 3l^3/256EI$$

$$a_{12} = a_{21} = 11l^3/768EI = a_{23} = a_{32}$$

$$a_{13} = a_{31} = 7l^3/768EI$$

$$a_{22} = l^3/48EI$$

When these values are substituted in equations 15.1, 15.2 and 15.3, and simplified

$$Y_1 = \frac{m\omega_n^2 l^3}{768EI}(9Y_1 + 44Y_2 + 14Y_3) \tag{15.4}$$

$$Y_2 = \frac{m\omega_n^2 l^3}{768EI}(11Y_1 + 64Y_2 + 22Y_3) \tag{15.5}$$

$$Y_3 = \frac{m\omega_n^2 l^3}{768EI}(7Y_1 + 44Y_2 + 18Y_3) \tag{15.6}$$

It is possible to eliminate the ratio of amplitudes and derive a frequency equation from which the natural frequencies of the three principal modes can be determined. However, this is a lengthy procedure and will not be attempted here. If only the lowest natural frequency is required it can be found by assuming a value for the amplitude ratio (they will all be in phase in the fundamental mode) and refining this by substitution into the right-hand sides of equation 15.4, 15.5 and 15.6.

It is left to the reader to verify that values of $Y_1:Y_2:Y_3 = 0.69:1:0.71$ fit these equations. Any of the three equations then gives

$$\omega_{n1} = 2.97 \sqrt{\frac{EI}{ml^3}} \quad \text{or} \quad f_{n1} = 0.473 \sqrt{\frac{EI}{ml^3}}$$

15.2.3 Uniformly Distributed Mass—Exact Solution

As before, let the vibration of the beam be represented by $y = Y \sin \omega_n t$, where Y is a function of x. If $\bar{m}$ is the mass per unit length, then figure 15.2 shows the inertia force of an element of length δx. Applying beam-deflection theory

$$EI \partial^4 y / \partial x^4 = \text{intensity of inertia loading}$$

$$= \bar{m} \omega_n^2 Y \sin \omega_n t$$

which gives

$$EI(\mathrm{d}^4 Y / \mathrm{d}x^4) = \bar{m} \omega_n^2 Y$$

The general solution can be written

$$Y = A \sin \alpha x + B \cos \alpha x + C \sinh \alpha x + D \cosh \alpha x \qquad (15.7)$$

where $\alpha^2 = \omega_n \sqrt{(\bar{m}/EI)}$, and the constants of integration can be found from the boundary conditions.

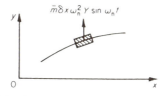

Figure 15.2   Distributed load

Example 15.2

Determining the natural frequencies of a simply supported beam of length l in terms of the mass per unit length $\bar{m}$ and the flexural rigidity EI.

The end conditions to be satisfied are that the deflection and bending moment are zero, that is $x = 0$, $Y = 0$ and $\mathrm{d}^2 Y / \mathrm{d}x^2 = 0$ and $x = l$, $Y = 0$ and $\mathrm{d}^2 Y / \mathrm{d}x^2 = 0$. From equation 15.7 the first two conditions show that $B = D = 0$, and the second two conditions that $C = 0$ and $A \sin \alpha l = 0$. Since A cannot also be zero if the beam is vibrating, it follows that $\sin \alpha l = 0$, that is

$$\alpha l = \pi, \ 2\pi, \ 3\pi \ \dots$$

The first three modes of vibration take up the shapes indicated in figure 15.3 and the corresponding natural circular frequencies are

$$\omega_n = \frac{\pi^2}{l^2}\sqrt{\frac{EI}{\bar{m}}}, \qquad \frac{4\pi^2}{l^2}\sqrt{\frac{EI}{\bar{m}}}, \qquad \frac{9\pi^2}{l^2}\sqrt{\frac{EI}{\bar{m}}}$$

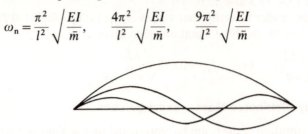

Figure 15.3 Modes of vibration

15.2.4 Dunkerley's Approximation to Fundamental Frequency

Dunkerley's method applies to the point mass case, and as such can be derived from the exact analysis of section 15.2.2.

Equations 15.1, 15.2 and 15.3 which refer to the system of figure 15.1 can be solved by elimination of the amplitudes to produce the determinant

$$\begin{vmatrix} a_{11}m\omega_n^2 - 1 & a_{12}4m\omega_n^2 & a_{13}2m\omega_n^2 \\ a_{21}m\omega_n^2 & a_{22}4m\omega_n^2 - 1 & a_{23}2m\omega_n^2 \\ a_{31}m\omega_n^2 & a_{32}4m\omega_n^2 & a_{33}2m\omega_n^2 - 1 \end{vmatrix} = 0$$

If this determinant is expanded, the resulting frequency equation is of the form

$$C_1\omega_n^6 + C_2\omega_n^4 + C_3\omega_n^2 + C_4 = 0 \tag{15.8}$$

where

$$C_3 = a_{11}m + a_{22}4m + a_{33}2m = (\delta_1 + \delta_2 + \delta_3)/g \tag{15.9}$$

$$C_4 = -1 \tag{15.10}$$

(δ_1 = static deflection due to gravity which the mass at point 1 would produce, etc.). But, if ω_{n1}^2, ω_{n2}^2 and ω_{n3}^2 are the roots of equation (15.8) this can be written

$$C_1(\omega_n^2 - \omega_{n1}^2)(\omega_n^2 - \omega_{n2}^2)(\omega_n^2 - \omega_{n3}^2) = 0$$

from which

$$C_3 = C_1(\omega_{n1}^2\omega_{n2}^2 + \omega_{n2}^2\omega_{n3}^2 + \omega_{n3}^2\omega_{n1}^2)$$

and

$$C_4 = -C_1\omega_{n1}^2\omega_{n2}^2\omega_{n3}^2$$

hence

$$\frac{C_3}{C_4} = -\left(\frac{1}{\omega_{n1}^2} + \frac{1}{\omega_{n2}^2} + \frac{1}{\omega_{n3}^2}\right)$$

$$\approx -1/\omega_{n1}^2 \qquad\qquad (15.11)$$

where ω_{n1} is the lowest natural frequency. By comparison of equations 15.9 and 15.10 with equation 15.11, and extending to any number of masses, Dunkerley's formula becomes

$$\omega_{n1} = \sqrt{\frac{g}{\Sigma\delta}} \qquad\qquad (15.12)$$

Example 15.3

Use Dunkerley's approximation to determine the fundamental frequency of the system shown in figure 15.1.

$$\delta_1 = \frac{mg(l/4)^2(3l/4)^2}{3EIl} = \frac{3mgl^3}{256EI}$$

$$\delta_2 = \frac{4mgl^3}{48EI} = \frac{mgl^3}{12EI}$$

$$\delta_3 = \frac{2mg(3l/4)^2(l/4)^2}{3EIl} = \frac{3mgl^3}{128EI}$$

$$\omega_{n1} = \sqrt{\frac{768EI}{ml^3(9 + 64 + 18)}}$$

$$= 2.905\sqrt{\frac{EI}{ml^3}}$$

$$f_{n1} = 0.462\sqrt{\frac{EI}{ml^3}}$$

This shows an error of approximately 2 per cent when compared with the exact value derived in example 15.1. It will also be seen that, due to the terms omitted from equation 15.11, *Dunkerley's method always gives a value less than the true value.*

15.2.5 Rayleigh's Approximation (Energy Method)

The principle expounded by Lord Rayleigh was that, provided a reasonable assumption was made for the vibrating form, an accurate estimate of the natural frequency could be obtained by considering the energy transfer between the mean and extreme positions. Unless the assumed form is exact (for example, a half sine-wave is correct for a uniformly distributed load with simply supported ends) this method applies additional constraints on the beam and hence the *value for natural frequency will always be too high* (as opposed to Dunkerley's formula).

If the vibration is expressed as $y = Y \sin \omega_n t$, then the maximum velocity (when passing through the mean position) is $\omega_n Y$, and all the energy is then in the kinetic form. Depending on whether distributed and/or point masses are to be taken into account

$$KE = \int \frac{1}{2} \bar{m} (\omega_n Y)^2 \, dx + \Sigma \frac{1}{2} m (\omega_n Y)^2 \qquad (15.13)$$

When passing through the extreme position all the energy is in the strain form, and may be written in terms of the bending moment M

$$SE = \int \frac{M^2}{2EI} \, dx = \int \frac{1}{2} EI (d^2 Y / dx^2)^2 \, dx \qquad (15.14)$$

This is a general expression that can be applied in all cases, but sometimes it is more convenient to use the static deflection curve as the vibrating form. With that assumption the strain energy would be equal to the work done if the loads were applied gradually from zero, so that

$$SE = \int \frac{1}{2} \bar{m} g Y \, dx + \Sigma \frac{1}{2} m g Y \qquad (15.15)$$

By equating equation 15.13 to either equation 15.14 or 15.15, and re-arranging, alternative expressions are obtained

$$\omega_n^2 = \frac{\int EI (d^2 Y / dx^2)^2 \, dx}{\int \bar{m} Y^2 \, dx + \Sigma m Y^2} \qquad (15.16)$$

$$\omega_n^2 = g \frac{\int \bar{m} Y \, dx + \Sigma m Y}{\int \bar{m} Y^2 \, dx + \Sigma m Y^2} \qquad (15.17)$$

Of these, equation 15.16 is usually simpler to use numerically, and has the added advantage of being applicable to higher modes if required.

Example 15.4

Obtain expressions for the fundamental frequency of lateral vibrations of a uniform beam of length l simply supported at each end, if the vibrating shape is assumed to be (a) $Y = A \sin \pi x / l$, (b) $Y = Cx(l - x)$, or (c) the static deflection curve.

Let $\bar{m}$ be the mass per unit length and EI the flexural rigidity.

(a) Equation 15.16 gives

$$\omega_n^2 = \frac{EI\pi^4 \displaystyle\int_0^l \sin^2(\pi x/l)\,dx}{\bar{m}l^4 \displaystyle\int_0^l \sin^2(\pi x/l)\,dx}$$

$$\omega_n = \frac{\pi^2}{l^2}\sqrt{\frac{EI}{\bar{m}}}$$

This is identical with the result in example 15.2 using the exact method, because the true vibrating shape has been used.

(b) Equation 15.16 gives

$$\omega_n^2 = \frac{EI \displaystyle\int_0^l 4\,dx}{\bar{m} \displaystyle\int_0^l x^2(l-x)^2\,dx}$$

$$\omega_n = \frac{10.95}{l^2}\sqrt{\frac{EI}{\bar{m}}}$$

This is nearly 11 per cent too high, the unusually large error arising from the fact that, although the parabola assumed has zero deflection at each end, it does not give zero bending moment ($M = EI\,d^2Y/dx^2$). So the condition for simply supported ends is not satisfied.

(c) The static deflection curve for a uniformly loaded beam is

$$Y = \frac{\bar{m}g}{24EI}(x^4 - 2lx^3 + l^3x)$$

Equation 15.17 becomes

$$\omega_n^2 = \frac{g \displaystyle\int_0^l \bar{m}Y\,dx}{\displaystyle\int_0^l \bar{m}Y^2\,dx}$$

$$= \frac{24EIl^5/5}{\bar{m} \times 31l^9/630}$$

$$\omega_n = \frac{9.88}{l^2}\sqrt{\frac{EI}{\bar{m}}}$$

This result is within 0.1 per cent of the true value.

Example 15.5

Obtain an approximate expression for the lowest natural frequency of the system shown in figure 15.1, assuming the beam vibrates in a half sine-wave.

Equation 15.16 must be used, and for three point loads only this reduces to

$$\omega_n^2 = \frac{EI \int_0^l (d^2 Y/dx^2)^2\, dx}{\sum_1^3 mY^2}$$

where Y_1 is the static deflection at m_1 which would be caused by all three loads $m_1 g$, $m_2 g$ and $m_3 g$ acting together at their respective positions along the beam. Similarly for Y_2 and Y_3. In this case $m_1 = m$, $m_2 = 4m$ and $m_3 = 2m$, and if the vibrating shape is assumed to be the half sine-wave $Y = A \sin \pi x/l$

$$Y_1 = A \sin \pi/4 = 0.707A$$

$$Y_2 = A \sin \pi/2 = A$$

$$Y_3 = A \sin 3\pi/4 = 0.707A$$

$$\omega_n^2 = \frac{EI\pi^4}{l^4} \frac{\int_0^l \sin^2(\pi x/l)\, dx}{m(0.707^2 + 4 + 2 \times 0.707^2)}$$

$$\omega_n = 2.976 \sqrt{\frac{EI}{ml^3}}$$

$$f_n = 0.474 \sqrt{\frac{EI}{ml^3}}$$

This result should be compared with the exact solution of example 15.1 and Dunkerley's approximation in example 15.3.

15.2.6   Note on Sign Convention for Deflections

The reader may have noticed that in all the examples so far dealt with, positive values have been assumed for deflections when substituting in the approximate formulae (and earlier for influence coefficients). It will be appreciated, however, that if the y-axis is taken vertically upwards static deflections will normally be negative, but since the static load is also negative the energy product mgY will remain positive. In fact, provided the deflection is in the same direction as the load the work done (and hence the strain energy stored) must be positive.

When using the method of influence coefficients the cross-coefficients may be negative (for example if there are overhanging loads), but since they only appear as squared terms in the frequency equation they may be treated as positive. Similarly in Dunkerley's approximation the deflections $\delta_1, \delta_2 \ldots$ are positive.

To sum up, whatever convention is used to calculate deflections, the 'modulus' value can always be used together with positive values for the masses.

15.2.7   Mixed Loading and Variable Beam Sections

If both distributed and point masses are to be allowed for, for instance when the mass of the beam cannot be neglected in relation to the loads carried, it is possible to replace the distributed mass by adding a proportion to each point mass. Bearing in mind that the aim is to have the same vibrational energy in the actual and equivalent systems, and that energy is proportional to the square of the amplitude, it will be found that the total added mass will generally be less than the actual distributed mass (some of which will be near a support). For instance, consider a simply supported beam having mass per unit length of $\bar{m}$. If this is to be replaced by a similar beam with a central point mass of m, then equating the natural frequency for the two cases

$$\frac{\pi^2}{l^2} \sqrt{\frac{EI}{\bar{m}}} = \sqrt{\frac{48EI}{ml^3}}$$

$$m = 0.493\bar{m}l$$

that is, the distributed mass is equivalent to just under half its total placed at the centre. Where a beam carries several point loads, these, together with the support points, divide it into a number of sections. An allowance for the mass of the beam can then be made by adding half the mass of each section to the load at each end of that section. In this way mass near to a support is rightly neglected.

When the beam section is not constant over its length, it is recommended that Rayleigh's method be used, based on the static deflection curve. However, except for symmetrical loading, this will involve an elaborate graphical or numerical procedure.

15.2.8   Overhanging Loads

If any loads are attached at points beyond the supports, as in figure 15.4, it is clear that, when vibrating, the inertia force on any overhung load will be in the opposite sense to that on a load between the supports. When using the energy method based on the static deflection curve it is necessary to assume that the weight acts *upwards* on the overhanging load. A numerical example will clarify this point.

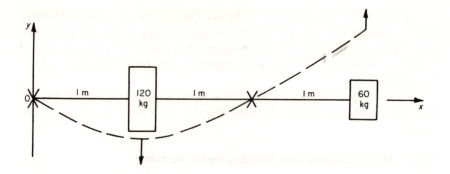

Figure 15.4 Overhanging load

Example 15.6

Make an estimate of the natural frequency of lateral vibrations of the beam
loaded as shown in figure 15.4. $EI = 10^4 \ \text{Nm}^2$.

(a) *Rayleigh's method.* Determine the static deflections due to a
downwards load of $120g$ N between the supports and an upwards load of
$60g$ N at the free end *acting together*. The support reactions are $90g$ upwards
and $30g$ downwards, and the bending-moment equation is

$$EI \, d^2 Y/dx^2 = 90gx - 120g[x - 1] - 30g[x - 2]$$

(using the Macaulay notation for the square brackets, see Ryder, *Strength
of Materials*).

Integrating twice and satisfying the boundary conditions $Y = 0$ at $x = 0$
and at $x = 2$

$$EIY/g = 15x^3 - 20[x - 1]^3 - 5[x - 2]^3 - 50x$$

at $x = 1$ m

$$Y_1 = -35g/EI$$

at $x = 3$ m

$$Y_2 = 90g/EI$$

Equation 15.17, treating all deflections as positive

$$\omega_n^2 = EI \frac{(120 \times 35 + 60 \times 90)}{120 \times 35^2 + 60 \times 90^2}$$

$$= 151.7$$

$$f_n = 1.96 \ \text{Hz}$$

(b) *Dunkerley's method.* The deflections are calculated for each gravity load *acting alone*, that is, for the 120*g* N

$$\delta_1 = \frac{120g \times 2^3}{48EI} = \frac{20g}{EI}$$

and for the 60*g* N load it can be shown that

$$\delta_2 = 60g/EI$$

Then

$$\omega_n^2 = \frac{g}{\delta_1 + \delta_2} = \frac{EI}{80}$$

$$f_n = 1.78 \text{ Hz}$$

The true value lies between these two results, and probably nearer to the Rayleigh value, say 1.9 Hz.

15.3 WHIRLING OF SHAFTS

Due to small imperfections during manufacture there will always be some eccentricity between the axis of a shaft and the bearing centre-line. As the speed of rotation is increased, inertia effects will tend to increase this eccentricity but will be resisted by the stiffness of the *shaft acting as a beam.* At certain speeds these two effects will be equally balanced, causing large deflections to build up as in a resonant condition. These are known as the *whirling speeds.* Because the inertia force is of the form $m\omega^2 Y$ the whirling speeds are the same as the natural circular frequency of lateral vibrations, and all the formulae previously developed can be applied. Self-aligning or 'short' bearings correspond to simply supported beam supports and 'long' bearings may be considered equivalent to the directional constraint of 'fixed' ends. In practice it is unlikely that either extreme condition will apply and the correct assessment is a matter of judgment.

Although these methods are adequate for the determination of whirling speeds they all lead to the conclusion that there will be no shaft deflection at any other speed. This in fact is not true, even if the effects of damping are neglected, and an examination of the behaviour of a shaft carrying a single rotor will show that the deflection varies with speed in the manner of a forced vibration.

15.3.1 Whirling of Single Mass Eccentrically Mounted

Let a mass *m* be attached to a rotating shaft such that its centre of mass is at a distance *e* from the shaft centre-line. If, when rotating at a speed ω, the shaft centre-line at the mass has a deflection y from the bearing axis, the inertia force is $m(y+e)\omega^2$ (note that, in the absence of damping, y and e must be in line but could be of opposite sign). This is resisted by the restoring force ky, where k is the stiffness of the shaft treated as a beam. Equating these forces

$$m(y+e)\omega^2 = ky$$

$$y = \frac{me\omega^2}{k - m\omega^2}$$

$$= \frac{\omega^2/\omega_n^2}{1 - \omega^2/\omega_n^2}e \qquad (15.18)$$

where $\omega_n = \sqrt{(k/m)}$ is the whirling speed.

For speeds below whirling y and e are in the same sense; for speeds above whirling they are in opposite senses. At very high speeds $y \to -e$ and the shaft rotates in a highly stable condition with the centre of mass on the bearing axis.

Example 15.7

A 15 mm diameter shaft rotates in long fixed bearings 0.6 m apart, and a disc of mass 20 kg is secured at mid-span. The mass centre of the disc is 0.5 mm from the shaft axis.

If the bending stress in the shaft is not to exceed 120 N/mm² find the range of speed over which the shaft must not run. $E = 200\,000$ N/mm².

The shaft must be treated as a beam fixed at both ends, for which the central deflection due to a load W is

$$y = Wl^3/192EI$$

and the corresponding maximum bending moment is $Wl/8$, giving a maximum bending stress

$$\sigma = \frac{Wl}{8} \times \frac{d/2}{I}$$

Hence the maximum value of y may be written in terms of the allowable stress

$$y = \frac{l^2\sigma}{12Ed}$$

$$= \frac{600^2 \times 120}{12 \times 200\,000 \times 15} = 1.2 \text{ mm}$$

But from equation 15.18

$$y = \frac{\omega^2/\omega_n^2}{1 - \omega^2/\omega_n^2} \times 0.5$$

where $y = \pm 1.2$ mm and

$$\omega_n^2 = 192EI/ml^3$$

$$= \frac{192 \times 200\,000 \times \pi \times 15^4}{20 \times 0.6^3 \times 64 \times 10^6}$$

$$= 22\,090$$

Substituting these values for y and ω_n^2 gives

$$\omega = 125 \text{ rad/s} = 1190 \text{ rev/min}$$

or

$$\omega = 195 \text{ rad/s} = 1860 \text{ rev/min}$$

The permitted stress will be exceeded if the shaft is run between these speeds.

SUMMARY

Lateral vibration of beams

Single mass on a light beam $\omega_n = \sqrt{(g/\delta)}$

Several masses on a light beam

Exact solution using influence coefficients

Dunkerley's approximation $\omega_{n1} = \sqrt{(g/\Sigma\delta)}$
Always underestimates ω_{n1}

Distributed mass

Exact solution for uniform distribution
using beam deflection theory

Rayleigh's approximation for both point
distributed masses

$$\omega_n^2 = \frac{\int EI(\mathrm{d}^2 Y/\mathrm{d}x^2)^2 \, \mathrm{d}x}{\int \bar{m}Y^2 \, \mathrm{d}x + \Sigma m Y^2}$$

or

$$\omega_n^2 = g\frac{\int \bar{m}Y \, \mathrm{d}x + \Sigma m Y}{\int \bar{m}Y^2 \, \mathrm{d}x + \Sigma m Y^2}$$

Whirling of shafts

15.4 PROBLEMS

1. A beam ABCD is supported at A and D and carries two equal loads of 3000 kg at B and C. The stiffness of the beam is such that the influence coefficients at B and C are $\alpha_{BB} = 0.076$ mm per kN, $\alpha_{BC} = \alpha_{CB} = 0.094$ mm per kN, $\alpha_{CC} = 0.14$ mm per kN.

   Determine the frequencies of lateral vibration.

   [6.38 and 31.1 Hz]

2. A uniform beam of length l and total mass $\bar{m}l$ is fixed in direction at one end and free at the other, so that it can vibrate as a cantilever. Show that the natural frequencies are given by the roots of the equation $\cos \alpha l \cosh \alpha l = -1$, where $\alpha^2 = \omega_n \sqrt{(\bar{m}/EI)}$.

   Hence determine the lowest natural frequency.

   $[0.56\sqrt{(EI/\bar{m}l^4)}]$

3. Show that, for the purpose of calculating the fundamental frequency of lateral vibrations of a cantilever beam, the distributed mass may be replaced by a point mass at the free end, and estimate its value.

   $[0.24\ \bar{m}l]$

4. A beam 6 m long and simply supported at each end carries a distributed load of 7000 kg/m extending from a point 1 m from one support to a point 2 m from the other support.

   Neglecting the mass of the beam, estimate the frequency of lateral vibrations by either (a) assuming the whole load acts its centre of mass, or (b) replacing by three point loads and applying Dunkerley's formula. $EI = 10^8$ Nm^2.

   [5.33, 5.73 Hz]

5. Obtain an expression for the lowest natural frequency of the beam shown in figure 15.1, assuming the vibrating form is the same as the static deflection curve.

   $[2.98\sqrt{(EI/ml^3)}]$

6. Use Rayleigh's method to determine the natural frequency of a cantilever beam of length l carrying a uniformly distributed load of intensity $\bar{m}$ if the vibrating form is assumed to be (a) $Y = A(1 - \cos \pi x/2l)$, or (b) the static deflection curve.

   What is the error in each case when compared with the exact value?

   $[0.583\sqrt{(EI/\bar{m}l^4)}, 0.562\sqrt{(EI/\bar{m}l^4)}$; 4 per cent, 0.4 per cent]

7. A uniform beam of flexural rigidity EI and total mass M over a length l, is fixed at each end and carries a point mass m at the centre. Using Rayleigh's principle and assuming the vibrating form is given by $y = a \sin^2(\pi x/l)$ determine the lowest frequency of lateral vibrations.

   $[6.28\sqrt{[EI/(3M + 8m)l^3]}]$

8. A cantilever ABC of length $5a$ is made from two lengths of tube, AB $= 2a$ and BC $= 3a$. The axes of the tubes are coincident and the end A is fixed. For the tube AB the second moment of area about a diameter is $2I$ and for the tube BC is I. The modulus of elasticity is E for both tubes.

   If the cantilever carries three masses $2M$, M, and M at distances from A of a, $3a$, and $5a$ respectively, show that the fundamental frequency of transverse vibration is approximately $0.03 \sqrt{(EI/Ma^3)}$.

9. A tubular steel shaft 1.5 m long having external and internal diameters of 35 mm and 25 mm is coupled through universal joints at each end. Evaluate the first and second whirling speeds if the density of steel is 7760 kg/m^3 and $E = 205\,000$ N/mm^2.
   [2315, 9260 rev/min]

10. A shaft 25 mm diameter is supported in self-aligning bearings 400 mm apart and carries a wheel of mass 50 kg at a point 150 mm from one bearing. If $E = 205\,000$ N/mm^2 what is the whirling speed?

   If the centre of mass of the wheel is 1.5 mm from the shaft axis, find the deflection of the shaft when running at 3000 rev/min.
   [2473 rev/min; 4.68 mm]

16 Robotics and Control

16.1 INTRODUCTION

As equipment and systems become more complex and sophisticated, the problems of operation and control by humans have become more difficult. Modern industrial processes and production techniques often require continuous vigilance of machine tools and equipment and entail boring repetitive jobs, where human error may lead to very expensive consequences, and put the operator, and possibly even the general public, at risk. In some cases, such as in the nuclear power industry and off-shore oil exploration and recovery, activities may need to be carried out in hazardous environments, and this may add ethical issues to the problems. Many of these difficulties and dangers can be relieved by the use of *robots*, devices which, properly designed, can operate faster, with greater accuracy and reliability, and with less fatigue, than their human counterparts, to give greater work output of a more consistent quality with increased safety.

Figure 16.1 shows a schematic example of a relatively simple device known as a *revolute* robot. In many ways this resembles part of a human, and some of its parts are described in human terms; it is particularly suitable in what might otherwise be a human workstation. A suitable rotary power actuator (for example, a stepper motor) is located at each of the six joints to control rotation about the specified axis so that the implement (for example, a paint spray gun) can be positioned and given the required motion. There must additionally be some control for the nozzle of the spray gun. It is clear from figure 16.1 that robots are *machines*, as defined in chapter 1, and some of the important aspects of their behaviour, for which much of the theory has already been covered, are listed below:

Kinematics Motion of the implement to perform the required operation without interference with itself, the workpiece, and adjacent equipment and personnel.

Kinetics Force, torque and power requirements, and inertia effects.

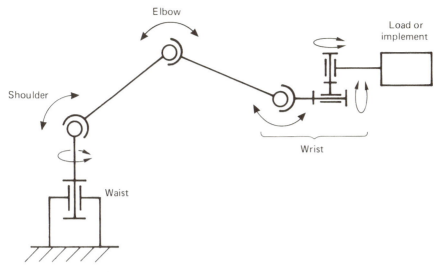

Figure 16.1   Revolute robot

Friction and Lubrication   Smooth jerk-free motion is particularly important where accurate positioning is required, such as for micro-electronics assembly work.

Vibration   Results from joint and arm flexibilities.

However, there are two important additional attributes which a machine must possess before it may properly be called a robot; namely, the ability to be programmed (and reprogrammed), and the ability to respond to its environment by means of an appropriate control system.

16.1.1   Robot Programming

Returning to the robot for paint spraying shown in figure 16.1, it is clear that the nozzle must be moved over the surface to be covered (which may be three dimensional) with the appropriate speed and stand-off distance, and in such a way that the whole surface is covered without wastage. Predetermined commands for the various joint actuators that describe the necessary motion are prepared and stored in a computer, generally one supplied specially for the purpose. In this particular application, a smooth motion requires simultaneous operation of several actuators, and thus some degree of parallel or distributed processing is necessary. In other applications, such as one where a robot is required to pick up a stationary object from one location and place it in another, and where there are no constraints on the path traced out by the robot, a sequence of commands to individual actuators may be quite acceptable. A change in the shape or size of the workpiece will clearly require the robot to be reprogrammed. Because of its specialised nature, detailed information on robot programming must be obtained from the manufacturer.

16.1.2 Robot Control

A robot programmed in the manner just described would have a very limited performance without some form of integrated control system. For example, a command to move to a particular position might not give an accurate result if there were significant backlash in the linkages. Again, returning to figure 16.1, the paint quality might be poor and inconsistent if the location of the workpiece was not very precise. Both of these risks could be overcome by using a *position control system* that could sense the actual positions, or relative positions, as appropriate, compare these with the required positions, and make the necessary corrections. A similar potential problem can arise where a controlled force is required, say, to pick up a fragile object. The gripping force must be sufficient to generate the required friction force, but not so great that the object is damaged. In some advanced systems, the robot program may be modified based on the errors which the control system detects, so that the robot is said to be *intelligent* and can *learn*.

Although particularly relevant to robots, control systems have a wider applicability, and the subject matter of this chapter is not restricted to robot applications. Initially it is useful to consider applications with which the reader is likely to have some familiarity, and where the 'machine' operator forms a vital part of the control system.

16.2 CONTROL SYSTEMS

It is common practice to represent control systems, whether automatic or human by means of *block diagrams*. Each block consists of some part of the system having a defined function, and is joined to other blocks by lines that represent a flow of information or material. The division of the system into blocks will depend entirely on the particular points of interest. Thus a complete engine and gearbox forming part of a control system may be represented by a single block, figure 16.2a, if all that is required is the relationship between the power output and the fuel input. In other cases it may be more convenient to represent it by a series of blocks as in figure 16.2b.

Control systems can be divided into two broad groups, *open-loop* and *closed-loop*. An open-loop system is one in which the output is related to the input through the basic design, and the relationship will remain constant only as long as the external conditions remain constant. Consider for example the speed control of a railway diesel locomotive. In figure 16.3a the driver's judgement is not involved, he merely sets the throttle to some position which

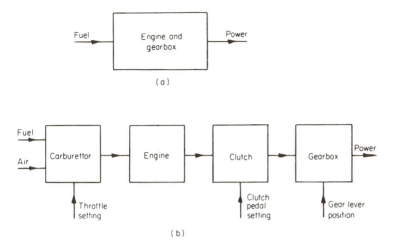

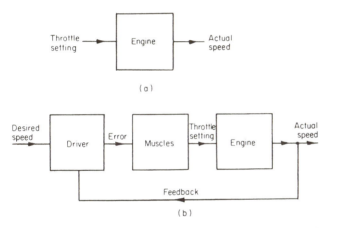

Figure 16.2   Block diagrams

Figure 16.3   Open and closed loops (a) open-loop system; (b) closed-loop system

by previous calibration is known to correspond to the desired speed. This open-loop system will be satisfactory while the track remains level, but gradients will vary the load, causing the speed to fluctuate (for the same throttle position), and if these fluctuations are unacceptable, *feedback* must be added to give a closed-loop system. In this example the driver himself will probably form the feedback loop as shown in figure 16.3b, but the essential features would not be altered if the driver were replaced by, say, an electro-mechanical system. An *error detector* (the driver) compares the output (actual speed) with the input (desired speed) to give an *error signal or deviation*. This will generally be a weak signal and will therefore need to be amplified (through the driver's muscles) before feeding to the engine in such a way that the error is reduced. In practice of course, the driver will probably respond to the way in which the error is changing, as well as to its magnitude, and in a similar manner, a fully automatic system may have quite complex feedback loops.

It should be clear already that a robot by itself can be thought of as an open-loop system.

16.3   TRANSFER FUNCTIONS

It is evident from the previous section that the relationship between the input and output is of prime importance in control systems. Such a relationship defines the response of the system and is known as a *transfer function*. It may refer to the complete system or to one or more individual blocks of the block diagram, and it should be noted that the input and output quantities may be of many different kinds and not necessarily of the same kind. Whatever their nature it is almost universal practice to represent them all by the symbol θ.

Almost all transfer functions are differential equations, and are usually written in the operator D notation. This has the advantage of being concise and simplifying the classification of control systems.

16.4   PROPORTIONAL CLOSED-LOOP CONTROL SYSTEMS

An automatic *position-control system* is shown schematically in figure 16.4a and in block-diagram form in figure 16.4b. The system has been shown as

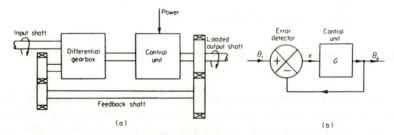

Figure 16.4   *Closed-loop position-control system*

essentially mechanical so that it is easier to visualise, but in practice modern control systems are largely electronic. The function of this system is to drive the loaded output shaft in such a way that it follows as closely as possible the movements of the command input shaft. For the error detector, a differential gearbox in this case

$$\varepsilon = \theta_i - \theta_o$$

The control unit produces an output proportional to the error so that

$$\theta_o = G\varepsilon$$

Eliminating ε, the transfer function or *closed-loop gain* is

$$\theta_o/\theta_i = G/(1 + G)$$

Now if the feedback loop is disconnected, the error is equal to the input and therefore the *open-loop gain* is simply G, hence

$$\frac{\theta_o}{\theta_i} = \frac{\text{open-loop gain}}{1 + \text{open-loop gain}} \qquad (16.1)$$

Figure 16.4 can in fact be used to represent a wide range of closed-loop control systems whose transfer functions will all be of the same form as equation 16.1. In terms of the paint spraying robot of figure 16.1, for example, the input signal θ_i, with its derivatives, represents the required nozzle motion to give the correct spray pattern, while the output signal represents the actual position of the nozzle. The control unit is the actuator (or actuators) used to move the nozzle.

It is clear that a *unity closed-loop gain* (giving output equal to input) will be approached if the *open-loop gain is made large*. For the position-control system this will be achieved if the control unit produces a very large effort per unit error, thus giving a high sensitivity. In practice, however, a high sensitivity will sometimes set up undesirable oscillations of the load, so that a compromise must be made.

16.4.1 First-order Position-control System

Consider a control unit that produces a torque proportional to the difference between the input and output displacements, the torque being applied to a load that gives a resistance increasing linearly with load velocity, figure 16.5a.

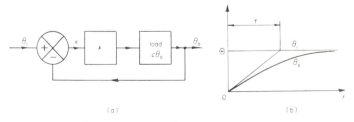

Figure 16.5 First-order position-control system

Such a load would occur when positioning a body of negligible mass in a viscous fluid. Equating the applied torque to the load torque

$$k(\theta_i - \theta_o) = c\dot{\theta}_o = cD\theta_o$$

hence

$$\theta_o/\theta_i = 1 \left/ \left(1 + \frac{c}{k}D\right)\right. = 1/(1 + \tau D) \tag{16.2}$$

where $\tau = c/k$.

A common method of comparing control systems is to examine their response to a step input, say $\theta_i = \Theta$. The combined transient and steady-state solution of equation 16.2 is then

$$\theta_o - \Theta(1 - e^{-t/\tau}) \tag{16.3}$$

and is shown in figure 16.5b. This response is often described as an *exponential lag* and it is clear that theoretically the output never quite reaches the input. Differentiating equation 16.3 and putting $t = 0$, the initial slope (output shaft velocity) is $\dot{\theta}_o = \Theta/\tau$. Thus if the output were to continue responding at the same rate, it would equal the input after a time τ. τ is defined as the *time constant* of the system. Comparing equations 16.1 and 16.2 it is seen that the open-loop gain is $1/\tau D$ and hence τ must be small for a good response.

Example 16.1

The device shown in figure 16.6 is known as a hydraulic relay. Initially, the system is at rest with the spool valve closing the ports to the lower power piston. When, say, a step input signal is supplied, corresponding to a movement θ_i to the right, the spool valve is also displaced to the right. This connects the high-pressure oil supply to the right-hand side of the power piston, which moves to the left to give the output θ_o. *At the same time* this leftward movement of the power system operates the lower end of the input linkage, returning the spool valve towards its original position.

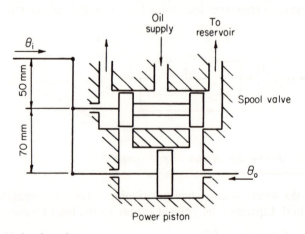

Figure 16.6   Hydraulic relay

Determine the time constant for a small step input θ_i assuming that the oil flow to the power piston is 550 mm^3/s per mm of spool valve displacement. The area of the power piston is 150 mm^2 and the masses of the moving parts may be neglected.

The spool valve is the error detector and its displacement is

$$\varepsilon = 7\theta_i/12 - 5\theta_o/12$$

Notice in this case that the error is *weighted*. Equating the oil flow through the spool valve to the product (piston area × velocity)

$$550(7\theta_i/12 - 5\theta_o/12) = 150\dot{\theta}_o = 150D\theta_o$$

hence

$$\theta_o/\theta_i = 1.4/(1 + 0.655D)$$

Comparing this with equation 16.2 the hydraulic relay is seen to be a first-order system and thus the time constant is 0.655 s.

16.4.2 Second-order Position-control System

Consider a system similar to that of figure 16.5a but assume that in addition the load has a finite mass so that inertia effects are significant. The torque equation now becomes

$$k(\theta_i - \theta_o) = c\dot{\theta}_o + I\ddot{\theta}_o$$

where I is the moment of inertia of the load referred to the output shaft. Re-arranging in operator D notation

$$\frac{\theta_o}{\theta_i} = \frac{k}{k + cD + ID^2} \tag{16.4}$$

Alternatively, keeping the dot notation

$$I\ddot{\theta}_o + c\dot{\theta}_o + k\theta_o = k\theta_i \tag{16.5}$$

If the substitutions $c/I = 2\zeta\omega_n$ and $k/I = \omega_n^2$ are made, then the left-hand side of equation 16.5 is seen to be of the same form as equation 13.18, the quantities k and c corresponding to the stiffness and damping of figure 13.5a. The damping may be an inherent part of the load, as for example with a ship's rudder blade, or it may be deliberately added for reasons that will be explained later. If the damping is sub-critical ($\zeta < 1$), then the transient solution of equation 16.5 will have a decaying oscillatory form as shown in chapter 13 by equation 13.20 and figure 13.5b. The steady-state solution which will be superimposed on this will of course depend on the form of the input. For robots, oscillation is generally undesirable and damping is usually added to make the response slightly super-critical (typically $1 < \zeta < 1.1$).

Three important cases will now be considered.

(a) *Step Input*, $\theta_i = \Theta$

For sub-critical damping ($\zeta < 1$), the complete solution to equation 16.5 can be shown to be

$$\theta_o = \Theta\left[1 - \frac{e^{-\zeta\omega_n t}}{\sqrt{(1-\zeta^2)}}\sin(\omega_n\sqrt{(1-\zeta^2)}t + \phi)\right] \tag{16.6}$$

where

$$\phi = \tan^{-1}\frac{\sqrt{(1-\zeta^2)}}{\zeta} \tag{16.7}$$

and is shown plotted for a value of ζ of about 0.2 in figure 16.7a. This rather low value has been chosen deliberately so that the detail of the shape of the response is easier to see. For super-critical damping ($\zeta > 1$), the complete solution of equation 16.5 is

$$\theta_o = \Theta\left[1 - \frac{1}{2\omega_n\sqrt{(\zeta^2-1)}}(\alpha_1 e^{-\alpha_2 t} - \alpha_2 e^{-\alpha_1 t})\right] \tag{16.8}$$

where

$$\alpha_1 = \zeta\omega_n + \omega_n\sqrt{(\zeta^2-1)} \tag{16.9}$$

and

$$\alpha_2 = \zeta\omega_n - \omega_n\sqrt{(\zeta^2-1)} \tag{16.10}$$

and this is shown plotted for a value of ζ of about 1.1 in figure 16.7b.

The design requirements of a second-order position-control system can be expressed in relation to the step input response in a number of ways.

First overshoot (sub-critical damping only) is measured from the steady-state position and equals $e^{-\zeta\pi/\sqrt{(1-\zeta^2)}}$ (compare logarithmic decrement, chapter 13). A reasonably high value of ζ is needed for a low overshoot.

Response time is the time taken for the output to settle to within a defined *tolerance band* of the input. For a given tolerance band, usually defined as a percentage of the steady-state value, there will be a value of ζ for which the response time is a minimum, and rather tedious manipulation of equations 16.6 and 16.8 will show that this always occurs with sub-critical damping where the first overshoot just reaches the upper limit of the tolerance band. Some typical values are given in table 16.1.

Rise time is the time taken for the output to rise from 10 per cent to 90 per cent of the input value. Increasing the value of ω_n will reduce the rise time for any given amount of damping.

It is clear, therefore, that the requirement for robot control systems to have sufficient damping to prevent oscillation is only achieved at the expense of longer response and rise times.

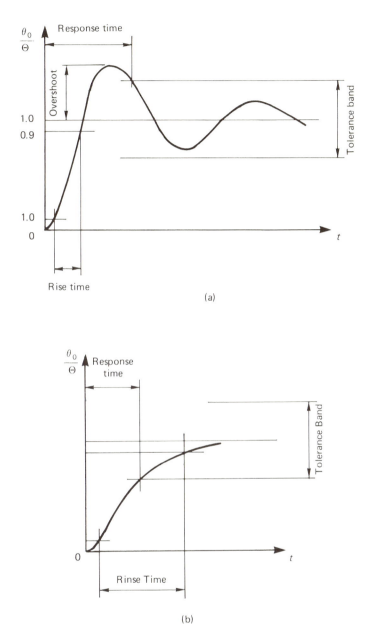

Figure 16.7 Step input response of a second-order position-control system
(a) sub-critical; (b) super critical

Table 16.1 Optimum Damping and Response Times

Tolerance band ± per cent	1	2	3	4	5	10	15	20
Optimum ζ	0.82	0.80	0.75	0.72	0.69	0.59	0.52	0.46
Response time × ω_n	4	3.5	3.2	3.0	2.8	2.3	2.0	1.8

(b) Ramp Input

This corresponds to a constant-velocity input at the command shaft, say ω_i, so that $\theta_i = \omega_i t$. When the input signal is first applied, the output will respond with the same transient as for case (a), according to the degree of damping. The steady-state output will be identical to the input, but will lag since there must be an error signal to produce the torque necessary to overcome the velocity component of the load. Thus, when the transient has died away, equation 16.4 gives

$$k(\theta_i - \theta_o) = c\dot{\theta}_o = c\omega_i$$

hence

$$\theta_o = \theta_i - c\omega_i/k = \theta_i - 2\zeta\omega_i/\omega_n \tag{16.11}$$

The term $2\zeta\omega_i/\omega_n$ represents the constant positional error or *velocity lag*, figure 16.8, and is clearly undesirable. The lag can be kept to a minimum by making ω_n large and ζ small, but the latter will result in a large first overshoot and a long response time to step input.

(c) Sinusoidal Input

The input will be of the form

$$\theta_i = \Theta_i \cos \omega t$$

Substituting into equation 16.5 and following the techniques for forced vibration described in chapter 13, the complete response will be

$$\theta_o = \text{transient solution} + \Theta_o \cos(\omega t - \beta) \tag{16.12}$$

where

$$\Theta_o = \frac{\Theta_i}{\sqrt{\left[\left(1 - \frac{\omega^2}{\omega_n^2}\right)^2 + 4\zeta^2\frac{\omega^2}{\omega_n^2}\right]}} \tag{16.13}$$

$$\tan \beta = \frac{2\zeta\omega/\omega_n}{1 - \omega^2/\omega_n^2} \tag{16.14}$$

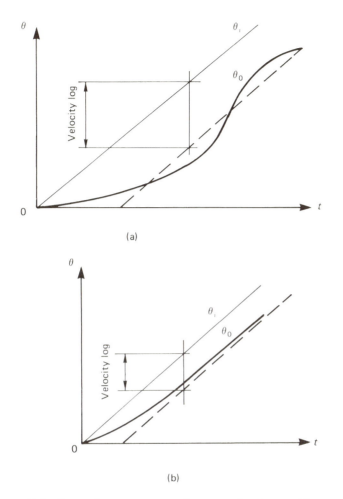

(a)

(b)

*Figure 16.8   Ramp input response of a second-order position-control system
(a) sub-critical; (b) super critical*

As with the vibration case, it is usually only the steady-state response that
is of interest with sinusoidal inputs, and thus the first term on the right-hand
side of equation 16.12 may normally be omitted. A plot of equations 16.13
and 16.14 will be the same as those shown in figures 13.8a and b, if the term
$(X/(F/k))$ is replaced by Θ_o/Θ_i, and it can be seen that a value of ζ around
0.67 will give a unity gain up to $\omega/\omega_n \approx 0.8$, but only if large phase-differences
are acceptable.

Example 16.2

A robot arm with error feedback is fitted with a damper that gives a value of $\zeta = 1.05$, and the rise time in response to a step input demand to the arm actuator is 0.8 s. Determine the value of ω_n and hence find the velocity lag when the arm is required to move at a steady angular velocity of 0.1 rad/s?

From equations 16.9 and 16.10

$$\alpha_1 = 1.37\,\omega_n \qquad \text{and} \qquad \alpha_2 = 0.73\,\omega_n$$

Substituting into equation 16.8 gives

$$\theta_o/\Theta = 1 - (2.141e^{-1.37\omega_n t} - 1.141e^{-0.73\omega_n t})$$

When $\theta_o/\Theta = 0.1$, trial solution of the above equation gives $\omega_n t_1 = 0.534$; and when $\theta_o/\Theta = 0.9$, $\omega_n t_2 = 4.131$.

But the rise time is

$$t_2 - t_1 = 0.8$$

therefore

$$\omega_n = (4.131 - 0.534)/0.8 = 4.50 \text{ rad/s}$$

and the velocity lag is

$$2\zeta\omega_i/\omega_n = 2 \times 1.05 \times 0.1/4.5 = 0.047 \text{ rad or } 2.67°$$

Example 16.3

An automatic control unit designed to position a flywheel produces a torque proportional to the error signal. The moment of inertia of the flywheel is 1.4 kg m^2 and its motion is viscously damped. When a small step is applied to the input, the first overshoot of the output is observed to be 10 per cent of the steady-state value, and when the input is driven at a steady speed of 2 rev/min, the output velocity lag is 3°. Determine the gain of the control unit.

If the input shaft is driven harmonically, over what range of frequency will the output shaft amplitude lie within ± 7 per cent of the input shaft amplitude?

For the step input

$$\text{first overshoot} = 0.10 = e^{-\zeta\pi/\sqrt{(1-\zeta^2)}}$$

giving

$$\zeta = 0.591$$

For the ramp input

$$\text{velocity lag} = 3 \times \pi/180 = 2 \times 0.591 \times \left(2 \times \frac{2\pi}{60}\right)/\omega_n$$

$$\omega_n = 4.73 \text{ rad/s}$$
$$= \sqrt{(k/I)}$$

hence control-unit gain is

$$k = 4.73^2 \times 1.4 = 3.13 \text{ Nm/rad}$$

The frequency response of the system will be given by equation 16.13 and thus

$$0.93 < \frac{\Theta_o}{\Theta_i} < 1.07$$

hence

$$\frac{1}{1.07^2} < \left(\frac{\omega}{\omega_n}\right)^4 + \left(\frac{\omega}{\omega_n}\right)^2 (4\zeta^2 - 2) + 1 < \frac{1}{0.93^2}$$

This solves to give only one real root

$$\omega/\omega_n > 0.893$$

hence

$$\omega > 4.23 \text{ rad/s} \qquad \text{or} \qquad f > 0.673 \text{ Hz}$$

Example 16.4

The angular position of a radar bowl is controlled by a second-order system that is required to give a minimum response time for a ± 5 per cent tolerance band. The velocity lag is not to exceed $2°$ per rev/min of the input command shaft. If the bowl has a moment of inertia of 108 kg m^2 and the gain of the control unit is 24.2 Nm per degree of error signal, what should be the viscous damping coefficient?

From table 16.1, the minimum response time for a ± 5 per cent tolerance band is achieved with a value of $\zeta = 0.69$. From equation 16.11, the velocity lag per unit input speed is

$$2\zeta/\omega_n < \left(2 \times \frac{\pi}{180}\right) \bigg/ \frac{2\pi}{60}$$

But

$$\omega_n = \sqrt{\left(\frac{24.2 \times 180}{108 \times \pi}\right)} = 3.58 \text{ rad/s}$$

hence

$$\zeta < \frac{2 \times 60 \times 3.58}{180 \times 2 \times 2} = 0.597$$

It therefore follows that the two requirements cannot be met simultaneously, and thus there must either be a compromise or a relaxation of the constraints.

16.5   CLOSED-LOOP CONTROL SYSTEMS WITH MODIFIED FEEDBACK AND FEEDFORWARD

It is evident from example 16.4, that a situation may arise where a proportional second-order system is incapable of meeting the design requirements. Even where the requirements can be met, the system has two major disadvantages arising from the viscous damping. First the damping will result in a dissipation of energy and therefore inefficiency, and second there will also be a velocity lag proportional to the amount of damping present. This section examines a number of modified feedbacks and the extent to which they overcome these disadvantages.

16.5.1   Negative-velocity Feedback

In this system a suitable transducer (for example a tachogenerator) produces a signal proportional to the output velocity which is then inverted and fed back to the controller, figure 16.9. The equation of motion is

$$k(\varepsilon - k_v \dot{\theta}_o) = c\dot{\theta}_o + I\ddot{\theta}_o$$

which can be rearranged to give the transfer function

$$\frac{\theta_o}{\theta_i} = \frac{k}{k + (kk_v + c)D + ID^2} \tag{16.15}$$

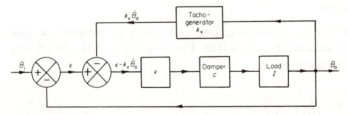

Figure 16.9   Negative velocity feedback

If equations 16.4 and 16.15 are compared, it is seen that the natural frequency is unchanged ($\omega_n = \sqrt{(k/I)}$) but the damping term has been increased by an amount $kk_v/I = k_v\omega_n^2$, therefore *increasing* the velocity lag. However, if the viscous damping can be reduced to keep the total effect the same, then the efficiency of the system is increased.

16.5.2   Positive-acceleration Feedback

In this case an accelerometer produces a signal proportional to the output shaft acceleration, which is then fed back positively in a similar manner to the negative velocity feedback of the previous system. Thus

$$k(\varepsilon + k_a\ddot{\theta}_o) = c\dot{\theta}_o + I\ddot{\theta}_o$$

where k_a is the accelerometer gain. Rearranging in the usual way

$$\frac{\theta_o}{\theta_i} = \frac{k}{k + cD + (I - kk_a)D^2} \tag{16.16}$$

It is now seen that acceleration feedback reduces the effective inertia of the load, thus increasing the natural frequency and giving a shorter rise-time. However, the system will be unstable if $I < kk_a$.

16.5.3   Error-rate Feedback

Here the error signal is passed through a modifying circuit which produces an output ($\varepsilon + k_1\dot{\varepsilon}$), so that the equation of motion becomes

$$k(\varepsilon + k_1\dot{\varepsilon}) = c\dot{\theta}_o + I\ddot{\theta}_o$$

and hence

$$\frac{\theta_o}{\theta_i} = \frac{k(1 + k_1D)}{k + (kk_1 + c)D + ID^2} \tag{16.17}$$

From this it is seen that the natural frequency is unchanged but the effective damping has been increased giving an improved transient response without the need for a high damping-torque acting on the output shaft. However, with a steady input velocity, the steady-state solution is

$$[k + (kk_1 + c)D]\theta_o = k(1 + k_1D)\theta_i$$

$$\theta_o = \theta_i - (c/k)\dot{\theta}_o = \theta_i - c\omega_i/k$$

so that the velocity lag is still determined only by the viscous damping c.

16.5.4   Input-rate Feedforward

In this system a velocity transducer on the input shaft produces a signal $k_2\dot{\theta}_1$ which is added to the error signal before passing to the control unit. The equation of motion is therefore

$$k(\varepsilon + k_2 \dot{\theta}_1) = c\dot{\theta}_o + I\ddot{\theta}_o$$

so that

$$\frac{\theta_o}{\theta_i} = \frac{k(1 + k_2 D)}{k + cD + ID^2} \tag{16.18}$$

From equation 16.18 the transient solution is unaffected, but for a steady input velocity ω_i, the velocity lag is now

$$\left(\frac{c}{k} - k_2\right)\omega_i = \left(\frac{2\zeta}{\omega_n} - k_2\right)\omega_i$$

and this lag can be made zero by correct choice of k_2.

16.5.5   Integral Control

An alternative method of eliminating the velocity lag is by the use of *integral control*. The signal fed to the control unit is then $k(\varepsilon + k_3 \int \varepsilon \, dt)$, and thus the equation of motion is

$$k(\varepsilon + k_3 \int \varepsilon \, dt) = c\dot{\theta}_o + I\ddot{\theta}_o$$

and hence

$$\frac{\theta_o}{\theta_i} = \frac{k(k_3 + D)}{(kk_3 + kD + cD^2 + ID^3)} \tag{16.19}$$

Substituting a steady input velocity ω_i into equation 16.19 shows that the lag is always zero; however, since a third-order equation is given, the possibility of instability must be investigated (see section 16.6).

Example 16.5

Find the steady-state velocity lag of the robot arm of example 16.2 if it is driven at a constant speed of 0.1 rad/s by an actuator that gives a torque of 54 Nm/rad of position error, against a constant friction torque of 8 Nm.

What input-rate feedforward constant should be used to reduce this lag to zero? Will the lag be zero at all speeds?

The equation of motion for the system is

$$k(\theta_i - \theta_o) = T + c\dot{\theta}_o + I\ddot{\theta}_o$$

where T is the friction torque and the other symbols are as previously defined. When the input demand is a constant speed, say ω_i, then

$$\theta_o = \theta_i - c\omega_i/k - T/k$$

so that the velocity lag is

$$c\omega_i/k + T/k = 2\zeta\omega_i/\omega_n + T/k = 0.047 + 8/54$$

$$= 0.195 \text{ rad} = 11.2°$$

With input-rate feedforward, the equation of motion is

$$k(\theta_i - \theta_o + k_2\dot{\theta}_i) = T + c\dot{\theta}_o + I\ddot{\theta}_o$$

and now the velocity lag is

$$-k_2\omega_i + c\omega_i/k + T/k$$

If this is to be zero when the speed is 0.1 rad/s then

$$k_2 = 2\zeta/\omega_n + T/k\omega_i = 2 \times 1.05/4.5 + 8/(54 \times 0.1)$$

$$= 1.948 \text{ rad per rad/s of input speed}$$

It is clear that, with the constant friction torque present, the condition of zero velocity lag can only be satisfied at one speed.

Example 16.6

A ship's rudder has a moment of inertia I and is subject to a retarding torque c times its angular velocity. It is positioned by a motor whose torque output is k times the input signal. During dock trials, this input signal is the position error, and the system is found to be satisfactory. When cruising, however, an additional centring torque acts on the rudder, equal to k' times its angular displacement. Explain how this will affect the response and determine the amount of any feedback or feedforward considered necessary.

The *in-dock* equation of motion is

$$k\varepsilon = c\dot{\theta}_o + I\ddot{\theta}_o$$

or

$$\frac{\theta_o}{\theta_i} = \frac{k}{(k + cD + ID^2)}$$

While *cruising*

$$k\varepsilon = k'\theta_o + c\dot{\theta}_o + I\ddot{\theta}_o$$

so that

$$\frac{\theta_o}{\theta_i} = \frac{k}{(k + k' + cD + ID^2)}$$

The additional term in the denominator has three effects,

 (i) to give a steady-state position error in response to a step input
(ii) to increase the natural frequency
(iii) to decrease the damping ratio

Consider these in order.

(i) From the cruising transfer function, a step input gives a steady-state solution

$$\frac{\Theta_o}{\Theta_i} = k/(k+k')$$

This can be restored to unity (zero error) by adding a *feedforward* term $(k'/k)\theta_i$ to give

$$\theta_o/\theta_i = (k+k')/(k+k'+cD+ID^2)$$

Notice that this modification will not affect the transient solution.

(ii) An increase in natural frequency will decrease the rise time and is therefore desirable. The response to a sinusoidal input will be poorer but this is unlikely to be of any consequence in this particular application.

(iii) The loss of viscous damping can be made by adding *negative-velocity feedback* with gain k_v so that

$$\frac{\theta_o}{\theta_i} = \frac{k+k'}{(k+k')+(kk_v+c)D+ID^2}$$

If this is to produce the same damping ratio as the *in-dock* situation, then

$$\frac{c}{2I}\sqrt{\frac{I}{k}} = \frac{c+kk_v}{2I}\sqrt{\frac{I}{k+k'}}$$

hence

$$k_v = \frac{c}{k}[\sqrt{(1+k'/k)}-1]$$

The velocity lag is now given by

$$\frac{\theta_o}{\theta_i} = \frac{k+k'}{k+k'+(kk_v+c)D}$$

hence

$$\varepsilon = \left(\frac{kk_v+c}{k+k'}\right)\omega_i$$

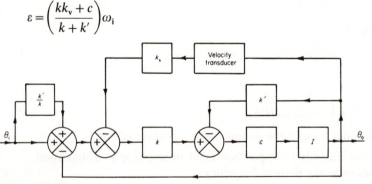

Figure 16.10

Substituting for k_v

$$\varepsilon = c\omega_i / \sqrt{(k(k + k'))}$$

This is an improvement over the *in-dock* velocity lag of $c\omega_i/k$. A block diagram of the modified system is shown in figure 16.10.

16.6 STABILITY

It was stated in section 16.5 that instability may arise with acceleration feedback and integral control, and indeed almost any control system may go unstable if improperly designed. This possibility is associated with the transient solution and appears as an oscillation of increasing amplitude following a disturbance. In practice, however, this increase will ultimately be limited by physical constraints or by the power available.

SUMMARY

Block diagrams—open- and closed-loop systems

Transfer functions

Proportional closed-loop position-control systems

First order $\quad \dfrac{\theta_o}{\theta_i} = 1 \left/ \left(1 + \dfrac{c}{k}D \right) \right.$

Second order

Error feedback $\quad \dfrac{\theta_o}{\theta_i} = \dfrac{k}{k + cD + ID^2}$

Velocity feedback $\quad \dfrac{\theta_o}{\theta_i} = \dfrac{k}{k + (kk_v + c)D + ID^2}$

Acceleration feedback $\quad \dfrac{\theta_o}{\theta_i} = \dfrac{k}{k + cD + (I - kk_a)D^2}$

Input rate feedforward $\quad \dfrac{\theta_o}{\theta_i} = \dfrac{k(1 + k_2 D)}{k + cD + ID^2}$

Integral control $\quad \dfrac{\theta_o}{\theta_i} = \dfrac{k(k_3 + D)}{kk_3 + kD + cD^2 + ID^3}$

Response to step, ramp and sinusoidal inputs

Stability

16.7   PROBLEMS

1. Figure 16.11 shows a form of hydraulic relay having a movable sleeve. The flow to the power ram of cross-sectional area A is Q times the valve opening. Derive the transfer function for the system and hence find the time constant.

$$\left[\theta_o/\theta_i = \frac{a}{b} \middle/ \left(1 + \frac{Aa}{Qb}D \right), \frac{Aa}{Qb} \right]$$

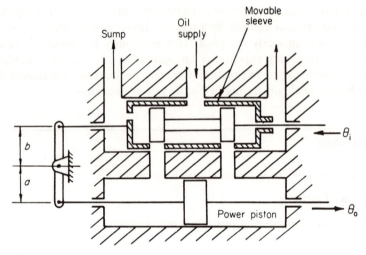

Figure 16.11

2. When a step input is applied to a second-order position-control system with error compensation, the first overshoot is measured and found to be just half of the steady-state response, and the natural frequency of decaying oscillations is 2.8 Hz. Determine the velocity lag in response to a steady rotational input of 25 rev/min.
   [6.84°]

3. Part of a robot is represented by a mass of 2000 kg positioned by a ram that produces a force proportional to error, the movement of the mass being viscously damped. The system is required to settle within a ±3 per cent tolerance band in a time of 2 seconds. Determine suitable values for the damping ratio and control unit constant.

   To reduce the velocity lag of the system to 0.4 mm per mm/s of input velocity without altering the transient response, some of the viscous damping is to be replaced by error-rate feedback. Determine the feedback constant.
   [5120 N/m, 0.538 m per m/s error rate]

4. A furnace door having a moment of inertia about its hinge axis of 5000 kg m² is driven through a 50:1 reduction gearbox by an electric

motor whose armature has a moment of inertia of 2 kg m^2. The motor torque/speed characteristics for a series of control winding supply voltages are given in figure 16.12. The error detector is fed directly from a position transducer on the load and an input command signal, and produces 10 V per radian of error to the control windings. Derive the transfer function for the door and hence find the equation for its displacement, after time t if the input signal is $\pi/8$ rad/s.

$[\theta_o/\theta_i = 4/(4 + 2D + D^2), \theta_o = \pi t/8 - (\pi/8\sqrt{3})e^{-t}\sin\sqrt{3}t]$

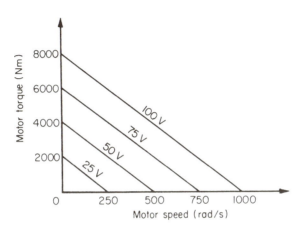

Figure 16.12

5. A profile milling machine consists of a cutter whose depth of cut is controlled by a hydraulic relay having a time constant of 0.1 s. The input to the spool valve of the relay is the difference between the position of a follower engaging a full-size template, and the position of the cutter. The template consists of a sine-wave of 14 mm wavelength. Determine the maximum feed rate if the steady-state output of the cutter is to be within 0.2 per cent of the follower movement.
    [1.41 mm/s]

6. A rotational inertia of 12 kg m^2 whose movement is subjected to viscous damping of 36 Nm per rad/s is to be positioned by a control unit producing a torque of 48 Nm per rad of error. To improve the transient response, the natural frequency is to be increased to 8 rad/s by the addition of acceleration feedback. Determine the constant of the feedback unit and the effect on the velocity lag.
    [0.234 Nm per rad/s^2, no change]

7. A simple robot for use in an automatic warehouse consists of a platform of mass 150 kg that can move up and down a vertical mast under the control of a viscous damper and a servo-motor that generates a force $F = (a + k\varepsilon)$ Newtons, where a and k are constants and ε is the platform height error in mm. The value of k is 300 N/mm of platform error. When

there is no payload on the platform, the height error is zero. What is the maximum payload that can be carried if the damping ratio ζ is not to fall below 1.15 nor to exceed 1.5, and what should be the value of the viscous damping rate c?

What will be the new maximum payload if output acceleration feedback is added, with a feedback constant equal to 80 per cent of the maximum allowable if there is to be no instability, and the damping rate remains unchanged?

[105.2 kg, 20.12 kN per m/s of platform velocity, 1126 kg]

INDEX